Perspektiven der Mathematikdidaktik

Reihe herausgegeben von

Gabriele Kaiser, Sektion 5, Universität Hamburg, Hamburg, Deutschland

In der Reihe werden Arbeiten zu aktuellen didaktischen Ansätzen zum Lehren und Lernen von Mathematik publiziert, die diese Felder empirisch untersuchen, qualitativ oder quantitativ orientiert. Die Publikationen sollen daher auch Antworten zu drängenden Fragen der Mathematikdidaktik und zu offenen Problemfeldern wie der Wirksamkeit der Lehrerausbildung oder der Implementierung von Innovationen im Mathematikunterricht anbieten. Damit leistet die Reihe einen Beitrag zur empirischen Fundierung der Mathematikdidaktik und zu sich daraus ergebenden Forschungsperspektiven.

Reihe herausgegeben von
Prof. Dr. Gabriele Kaiser
Universität Hamburg

Weitere Bände in der Reihe http://www.springer.com/series/12189

Macarena Larrain Jory

Preservice Primary Teachers' Diagnostic Competences in Mathematics

Assessment and Development

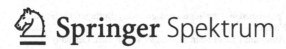

Macarena Larrain Jory
University of Hamburg, Hamburg, Germany
Universidad de los Andes, Santiago, Chile

Dissertation, Universität Hamburg, 2020

ISSN 2522-0799 ISSN 2522-0802 (electronic)
Perspektiven der Mathematikdidaktik
ISBN 978-3-658-33823-7 ISBN 978-3-658-33824-4 (eBook)
https://doi.org/10.1007/978-3-658-33824-4

Responsible Editor: Marija Kojic
This Springer Spektrum imprint is published by the registered company Springer Fachmedien
Wiesbaden GmbH part of Springer Nature.
The registered company address is: Abraham-Lincoln-Str. 46, 65189 Wiesbaden, Germany

Preface

This thesis explores the initial education of primary teachers in the mathematics area. In this vast field, the author focuses on the diagnostic competence of preservice teachers in error situations and finds very interesting results regarding this competence and its relations with variables associated with the background and disposition of preservice teachers. Moreover, the author designs a short seminar sequence to improve the diagnostic competence of preservice teachers in error situations and reports changes in this competence and studies the relations of these changes with the same variables mentioned above.

The analysis and conclusions of this study are drawn on research data from Chilean preservice teachers studying in various universities in the capital Santiago. However, the results shed light on general issues regarding the diagnostic competence in error situations, its relation with other variables and the possibility of developing it during teachers' initial education. In addition, this thesis opens very interesting research questions to further studying this crucial competence in mathematics education.

Rather than comment on the range of validity of the thesis' conclusions and its value in terms of the advance of research in the area, which is the role of thesis readers and future peer reviewers, I would like to discuss the value of this thesis regarding ideas, clues and consequences for the initial education of teachers in Chile. By doing so, readers from other countries may possibly identify some of the ideas that can be applied to their own reality.

Around the globe, changes and reforms in education and mathematics education have been taking place aiming to adjust goals and methods to the new realities imposed by technological changes. Chile is no exception, and national curricular changes have been taking place for about ten years. However, changes are slow at the school level and, paradoxically, even slower at the university level.

This situation can be observed in various ways, but data from this thesis may provide a different angle. As a result of the study, more than one-third of the sample did show instructivist preferences for dealing with student errors and another one third didn't show substantial answer. Thus, only one-third of preservice teachers in the sample were able to provide a constructivist approach to students to deal with the observed error, that is a response aligned with reform ideas.

As a conclusion of the thesis, a higher competence to hypothesize about causes of student errors is related to stronger mathematical knowledge for teaching, to practical experience in teaching and to constructivist beliefs about the nature of mathematics and about mathematics teaching and learning. Even more, constructivist beliefs are associated with the group of preservice teachers showing constructivist preferences for dealing with student errors, and teacher-directed beliefs about mathematics learning and views of mathematics as a set of rules and procedures are associated with the group of preservice teachers showing instructivist preferences for dealing with student errors. Considering this, we can suggest that initial teacher education programs should include a strong emphasis on mathematics for teaching and practice-related activities, i.e. real and/or simulated teaching. These conclusions allow us to be optimistic about the future since these activities are strongly encouraged in the initial education of teachers in Chile.

Interestingly enough, this thesis additionally provides a short seminar sequence that showed to have positive effects in improving the competence to hypothesize about causes of student errors and a slight increase of the number of preservice teachers showing constructivist preferences for dealing with those errors. Here again, the positive changes were related to mathematical knowledge for teaching and to practical experience. These results show the suitability of this short seminar setting, with a practical approach together with discussion and reflection activities, for improving the diagnostic competence during initial education, and possibly other teaching competences.

As a result of the analysis, the conception of mathematics and mathematics teaching and learning appears as a very important variable related to preservice teachers' diagnostic competence in error situations and to the changes after the short seminar sequence in which they participated. On the one hand, we have constructivist beliefs about the nature of mathematics, in which mathematics are viewed as a flexible construction process, were rules appear as necessary, and helps to understand and solve problems. On the other hand, the conception of mathematics as a set of fixed rules and procedures that have to be learned with related conceptions of teaching and learning mathematics. Can these conceptions be changed during teachers' initial education? Certainly, it is not something that a short seminar will achieve, but an active and constructive methodology in mathematics and mathematics teaching courses, aligned with reform ideas, should undoubtedly help in this direction. In this way, this thesis' results provide more reasons to be optimistic in the future of mathematics education.

Besides the practical consequences and ideas that this thesis provides for the development of teachers' initial education in Chile, it opens a great number of interesting questions to pursue research in diagnostic competence in error situations. Errors are definitely a rich source of teaching and learning opportunities in classrooms, as it is well documented in this thesis. I wonder why correct steps, answers or arguments are not considered as a source of teaching and learning opportunities for preservice teachers, allowing them to hypothesize about the reasons for the correct answers and about how to deal with them in order to improve students' understanding of mathematics.

I would like to end these words by congratulating Macarena Larrain for this well-grounded thesis, where she could answer several questions and open another set of interesting ones. I enjoyed and learned reading the thesis and it brought to me ideas and practical consequences for education in Chile.

Dr. Patricio Felmer
Professor
Universidad de Chile
Santiago, Chile

Acknowledgements

There were many who helped and accompanied me on this long journey. This is the moment to thank them and to rejoice all together.

First, I would like to deeply thank Prof. Dr. Gabriele Kaiser, whose insight and knowledge of the subject guided me through this research. Her support, patience, valuable advice and constant encouragement to achieve goals that would move me forward were crucial to successfully completing this research. Thanks also to Prof. Dr. Marianne Nolte for sharing her knowledge in the field of primary education, for the interesting discussions and conversations and for advising me and supporting me in this and other personal issues.

Further, I would like to thank the members of the research colloquium of the University of Hamburg for the enriching and thoughtful discussions about my research, which helped me on several occasions to look at things from new points of view. I would especially like to thank Dr. Hannah Heinrichs, who patiently and unconditionally helped me in various aspects of methodology and analysis of results. The same goes for Dr. Armin Jentsch and Dr. Nils Buchholtz, who were always willing to give thoughtful advice to my work. I am also grateful for the valuable advice that Dr. Katrin Vorhölter gave me at the colloquia and for having agreed to participate in the evaluation committee.

From the beginning of my doctorate, the support of Universidad de los Andes played a crucial role, for which I am extremely grateful. I would especially like to thank Dr. Adela López, Dr. Pelusa Orellana, Francisca Valenzuela and Antonieta Ramacciotti for their support and encouraging conversations.

I would also like to thank the many people who were essential to the instrument's development and the implementation of the seminar sequence. First, I would like to thank Daniela Ramírez, Alejandra Meneses and all the children and their parents, who helped with the production of the video vignettes in the Deutsche Schule Santiago. In addition, the support of Dr. Leonor Varas was crucial to having access to appropriate assessment instruments. Another immense gratitude is for the teacher educators for opening the doors of their institutions and their courses to me: Dr. Patricio Felmer, Alejandra Besa, Anita López, Dr. Roberto Vidal and Dr. Sebastián Howard (†). The student teachers who participated in the sessions and answered the questionnaires at the various institutions should also receive special thanks, not only because they allowed me to collect the data enabling this research, but because they made this an entertaining and very enriching experience.

And my biggest thanks to my family and friends for all the support you have shown me through this process. To my parents, thank you for always being present, for teaching me that I can achieve whatever I set out to do and for encouraging me to keep going. To my kids, thanks for being patient when I had to spend endless hours on my computer and sorry for being even grumpier than usual while writing this thesis! And for my husband, Kris, thanks for all your support, without it I would have certainly stopped these studies a long time ago. You have been amazing!

Introduction

Results from international assessments such as the Programme for International Student Assessment (PISA) or the Trends in International Mathematics and Science Study (TIMSS) have made evident the need to actively work towards improving the quality of school education in many countries. Additionally, 'concerns have been raised about whether preservice and in-service training succeeds in equipping teachers with the professional knowledge they need to deliver consistently high-quality instruction' (Baumert et al,, 2010, p. 133). In other words, advancing school effectiveness requires examining and improving teachers' competences and the learning opportunities provided by initial and further teacher education programs.

Most educational reforms across the globe demand that mathematics teaching shifts from a traditional teacher-directed instructional approach to the formation of competence-oriented teaching environments, in which student thinking is highlighted and taken as a starting point for building further mathematical knowledge (e.g., National Council of Teachers of Mathematics, 2000). Moreover, the increasing heterogeneity in classrooms sets new challenges for teachers, who have to be able to support students individually. Such an educational paradigm requires teachers equipped with a series of complex professional competencies that allow them to plan and carry out teaching strategies that consider all students' needs and provide sufficient and targeted learning opportunities. In particular, understanding students' thinking has been regarded as crucial to successfully differentiate learning experiences and use individual learner's reasoning as the basis on which to build further mathematical knowledge (Cooper, 2009; Empson, 2003; Jacobs, Lamb & Philipp, 2010).

In the field of mathematics education, the professional competencies needed by teachers have been extensively described, researched, and debated (see, for instance, Ball, Thames & Phelps, 2008; Blömeke, Gustafsson & Shavelson, 2015; Depaepe, Verschaffel & Kelchtermans, 2013; Kaiser, Blömeke, König, Busse, Döhrmann & Hoth, 2017; Shulman, 1986). It has been argued for the relevance of teachers' abilities to identify and understand what students know, what they still need to learn, and what they have incorrectly understood, to conduct ongoing analyses of students' learning and to make instructional decisions that support and cognitively activate student learning. Particularly significant have been research and discussions about teachers' noticing skills (Sherin, Jacobs & Philipp, 2011). This line of research has highlighted teachers' abilities to identify noteworthy classroom events and children's mathematical thinking and to interpret these observations by connecting them with prior knowledge and experiences (Jacobs et al., 2010; van Es & Sherin, 2002). A third component has been included, namely, teachers' skills to decide how to provide an instructional response based on what they have attended to and interpreted (Jacobs et al., 2010).

From another perspective, teachers' diagnostic competence has been regarded as essential for the evaluation and understanding of student thinking and, thus, for adapting and individualizing teaching strategies in a way that fosters further learning. According to Weinert, Schrader and Helmke (1990), diagnostic competence is one of the four key components of teacher expertise, together with classroom management competence, knowledge of instructional techniques, and subject-matter knowledge. Helmke (2017) justifies the particular importance that teachers' diagnostic competence has for the teaching and learning process because of its essential role in designing effective teaching strategies and aligning instructional responses to students' learning requirements. Prediger (2010) emphasizes the connection of teachers' diagnostic competence to a student-centered teaching style and argues that it is necessary to understand student thinking and take it as a starting point to build further learning and provide sensitive support to individual students and the whole class.

Additionally, it has been widely recognized that learning situations in which errors arise are very often a rich source of information for teachers (Ashlock, 2010; Prediger & Wittmann, 2009; Rach, Ufer & Heinze, 2013; Radatz, 1979; Scherer & Moser Opitz, 2012). Analysis of students' errors can uncover their erroneous conceptualizations or misconceptions and provide teachers with valuable insights into individual students' understanding of mathematical concepts and procedures. Based on this information, teaching strategies can be targeted to students' particular needs and, thus, better promote further mathematical learning.

Despite the agreement on the relevance of teachers' diagnostic competence for promoting student mathematical learning, more evidence is needed about the development of this competence and how it can be productively fostered in teacher education. In particular, preservice primary school teachers' diagnostic competence in error situations requires further attention in order to better understand key characteristics and the structure of the competence and to uncover critical aspects influencing its development. It has been suggested that teaching experience is a relevant factor for the development of this competence. However, it has also been argued that preservice teachers can benefit from guided experiences and knowledge that may constitute the basis for further developing this competence (Cooper, 2009; Heinrichs, 2015). Hence, it is the interest of this study to examine the characteristics of preservice primary school teachers' diagnostic competence in error situations and how it can be fostered. Consequently, the central research question of the present study is

To what extent is it possible to promote preservice primary school teachers' diagnostic competence in error situations within initial teacher education?

Towards that goal, a brief university seminar sequence was developed to be included in initial teacher education programs at Chilean universities. The seminar sequence introduces preservice teachers into the value of analyzing students' errors for improving teaching and aims at building the foundations for the development of their diagnostic competence. To investigate the effect of the seminar sequence, a computer-based pre- and post-test was developed to assess preservice teachers' diagnostic competence in error situations and evaluate their changes after taking part in the seminar sequence. The design of both the seminar sequence and the diagnostic-competence test was grounded on a study carried out in German universities with preservice secondary mathematics teachers (Heinrichs, 2015) and used its model of the diagnostic process in error situations, which consists of three facets: perceiving the error, hypothesizing about causes for the error and dealing with the error. The model's facets were also used in the present study to structure the characterization of preservice primary school teachers' diagnostic competence in error situations and describe its changes after participation in the seminar sequence.

To specify the construct of diagnostic competence in error situations and frame the interpretation and discussion of the results of the present study, the first chapter provides an overview of the current state of research on the field. The concept of teachers' professional competencies, as well as various models specifying it for mathematics teachers, are described and explained. This is followed by a review of different conceptualizations and models of the construct of teachers' diagnostic competence and its connection with teachers' professional competencies. Moreover, the relevance of teachers' diagnostic competence for the teaching and learning process and how it can be fostered within teacher education are described. Additionally, the role of errors in mathematics teaching and learning is revised and connected to the specification of the concept of diagnostic competence for learning situations in which errors arise. Finally, the three facets of the diagnostic process in error situations are described in detail as used in this study.

After presenting the guiding research question and hypotheses in the second chapter, the third chapter specifies the study's methodological context. It begins by displaying the study's design and detailed descriptions of the four sessions of the university seminar sequence and the pre- and post-test developed to assess preservice teachers' diagnostic competence in error situations. Next, a brief overview of data collection in Chilean universities is provided. Lastly, both the qualitative and quantitative methods used in the present study are presented. Thereupon, the method of qualitative text analysis is described, and, in particular, the procedures used to conduct evaluative qualitative text analysis in the present study are specified. In a similar way, the quantitative methods used in the study are described, namely Item-Response-Theory, Latent-Class-Analysis, and several statistical tests for hypotheses testing.

The fourth chapter presents the results of the conducted analyses. First, the results of the cross-sectional analyses of the data are reported, which allows describing preservice teachers' diagnostic competence in error situations and relating these characteristics with other features of preservice teachers' background. Similarly, the results of the longitudinal analyses of the data are presented and discussed. This allows examining the changes observed in preservice teachers' diagnostic competence in error situations after their participation in the university seminar sequence. Moreover, the relations of these changes with features of preservice teachers' backgrounds provide valuable indications about relevant aspects for promoting the development of preservice teachers' diagnostic competence within initial teacher education.

In the closing chapter, the main results of the study are summarized and discussed. Then, limitations of the present study, as well as arising questions, are explicated. The chapter ends by considering and discussing implications of the results for teacher education and further research opportunities.

Contents

List of Figures

List of Tables

State of the Art and Theoretical Framework

1.1 Teachers' Professional Competencies

In the light of unsatisfactory students' results in large-scale international assessments, great concern about the effectiveness of teachers' work has been raised, particularly in many Western countries. Closely together, criticism of teacher education in general has notably increased, which has, in turn, directed the attention and efforts of many researchers towards the study of teachers' professional competences, their nature and development. In particular, the professional competencies of mathematics teachers have gained much attention in the last decades.

1.1.1 The Construct of Professional Competence

Before going into details about the nature and characteristics of teachers' competencies, the concept of competence itself needs to be clarified. In educational research, the approach proposed by Weinert (2001) has strongly influenced the understanding of the concept of competence. He stated that it "refers to the necessary prerequisites available to an individual or a group of individuals for successfully meeting complex demands" and argued that these prerequisites "are comprised of cognitive and (in many cases) motivational, ethical, volitional, and/or social components" (Weinert, 2001, p. 62). Moreover, he pointed out that these prerequisites are acquired through learning. Hence, according to this approach, competencies are composed of a set of cognitive abilities, knowledge, skills and associated attitudes, motivational, volitional and social variables that

© The Author(s), under exclusive license to Springer Fachmedien Wiesbaden GmbH,
part of Springer Nature 2021
M. Larrain Jory, *Preservice Primary Teachers' Diagnostic Competences in Mathematics*, Perspektiven der Mathematikdidaktik,
https://doi.org/10.1007/978-3-658-33824-4_1

are learned by individuals and then available for successfully solving complex demands at specific situations. Furthermore, Koeppen, Hartig, Klieme and Leutner (2008) highlight the context-specificity of competencies. They suggest that competencies are connected to the specific domains in which tasks are to be solved and, therefore, they are to be developed through learning opportunities in situations relevant to the domain.

1.1.2   Initial Models of Teachers' Professional Competencies

Weinert's (2001) conceptualization of competence distinguishes between cognitive and affective or motivational aspects that are needed to meet the demands of complex tasks. Discussions about the competencies needed by mathematics teachers have largely focused on cognitive aspects. Various researchers have studied the nature of the knowledge needed by teachers to be effective in their occupation, many of them starting from the work of Shulman (1986), who introduced the concept of 'pedagogical content knowledge' (PCK) and largely influenced the field.

1.1.2.1 The Seminal Contribution of Lee Shulman
In his groundbreaking work, Shulman (1987) outlined seven categories into which the knowledge needed by teachers to promote learning can be structured. These included content knowledge, general pedagogical knowledge, curricular knowledge, pedagogical content knowledge, knowledge of students and their characteristics, knowledge of the educational context and knowledge about the purposes and values of education. He stressed the relevance of pedagogical content knowledge (PCK) because it is distinctive of the work of teaching, a "special form of professional understanding" (p. 8). He characterized it as "the blending of content and pedagogy into an understanding of how particular topics, problems, or issues are organized, represented, and adapted to the diverse interests and abilities of learners, and presented for instruction" (p. 8).

Shulman (1986) detailed three of the seven categories, namely on those related to content knowledge: subject matter content knowledge, pedagogical content knowledge and curricular knowledge. Regarding subject matter content knowledge, he emphasized that teachers need not only to know the facts and contents but also to understand the structures of the subject. Besides understanding concepts and knowing facts, teachers "must also be able to explain why a particular proposition is deemed warranted, why it is worth knowing, and how it relates to other propositions, both within the discipline and without, both in theory and in practice" (p. 9). The second category, pedagogical content knowledge, refers to the body of knowledge that enables teachers to teach the content to others. It includes, for the taught topics, a wide range of representations, powerful examples and various ways of explaining, knowledge about specific aspects of a content that students may find it easier or more difficult to understand, as well as common preconceptions and misconceptions related to the topic and strategies that may prove useful to support students in overcoming them. Thirdly, he referred to curricular knowledge as the understanding of the existing curricular strategies for teaching a topic at a certain grade level and knowledge about available teaching programs and instructional materials. Knowledge about topics taught in prior and following grade levels in the same subject and in parallel in other subjects is also part of teachers' curricular knowledge.

1.1.2.2   Mathematics Teaching and Learning to Teach (MTLT) and Learning Mathematics for Teaching (LMT) Projects

In order to develop more specific and precise definitions of the concepts of pedagogical content knowledge and subject matter knowledge that would allow the operationalization and measurement of these constructs in empirical research, a group of researchers at the University of Michigan analyzed primary school teachers' work (Ball & Bass, 2003). Within the project Mathematics Teaching and Learning to Teach (MTLT), they investigated what constitutes the knowledge related to mathematics that teachers need in their professional activities. To do this, they focused on the demands of teachers' daily tasks and analyzed their underlying mathematical requirements in terms of reasoning, understandings and skills. Their observations suggest that "the mathematical demands of teaching are substantial. The mathematical knowledge needed for teaching is not less than that needed by other adults. In fact, knowledge for teaching must be detailed in ways unnecessary for everyday functioning" (Ball et al., 2008, p. 396). In particular, they found that teachers perform a wide range of tasks that involved both mathematics and pedagogical demands, not only while teaching a lesson and interacting with students, but also when planning lessons, evaluating and grading students' work, managing homework and communicating with parents and other education professionals. Teaching mathematics in primary school classrooms involves knowing and choosing appropriate representations, examples and explanations to clarify concepts and procedures, examining, listening and understanding students' work, among a wide variety of tasks that involve both mathematical and pedagogical thinking (Ball, Hill & Bass, 2005). Based on these analyses, they developed a theoretical approach—Mathematical Knowledge for Teaching (MKT) –, which they defined as "a kind of professional knowledge of mathematics different from that demanded by other mathematically intensive occupations" (p. 17). In a subsequent study, the Learning Mathematics for Teaching (LMT) project, they developed assessment instruments to measure the content knowledge needed for teaching mathematics (Hill, Schilling & Ball, 2004). These instruments allowed studying the nature of their MKT construct, empirically identifying some subdomains and uncovering others.

According to Ball et al. (2008), MKT comprises both the subject matter knowledge and the pedagogical content knowledge facets described by Shulman (1987). The former is subdivided into common content knowledge (CCK), specialized content knowledge (SCK) and horizon content knowledge. The first dimension, common content knowledge (CCK), refers to the "mathematical knowledge and skill used in settings other than teaching" (p. 399), to the mathematics that teachers need to know that is not unique to the work of teaching. Besides knowing and understanding the topics in the curriculum, teachers need common knowledge to distinguish between students' correct and wrong answers and between accurate and inaccurate definitions and arguments. They have to know how to use concepts and notations mathematically correctly themselves and to identify when they are used incorrectly in textbooks or by students. The second dimension, specialized content knowledge (SCK), is "the mathematical knowledge and skill unique to teaching" (p. 400) and covers ways of understanding and unpacking mathematics that allow them to make the subject comprehensible for others. This goes beyond the contents taught to students and is also distinct to the knowledge needed in other settings or professions. For instance, it includes an understanding of mathematics that allows them to decide whether an unconventional method used by a student is mathematically correct, to identify and exemplify different interpretations of the operations, or to effectively select and use representations of mathematical concepts. The third category, which they called horizon knowledge and positioned provisionally into subject matter knowledge, refers to an awareness of the location and relations of mathematical contents in the curriculum, so teachers can build mathematics foundations for topics that come later.

The facet of pedagogical content knowledge is related to Shulman's conceptualization of PCK. It is composed of three dimensions, namely knowledge of content and students (KCS), knowledge of content and teaching (KCT) and knowledge of content and curriculum. The first dimension, KCS, focuses on the connection between teachers' understanding of how students learn and particular mathematics topics. It includes, for example, the ability to foresee aspects of the content that students may find difficult to grasp, some typical students' reasoning about a concept, or the answers that students are likely to give to certain tasks. Understanding students' thinking and arguments, even if they are expressed in incomplete or imprecise language, and knowledge about the most common errors and misconceptions for the topics being taught, are also examples of KCS as they require an interaction of knowledge about the subject and about students. The second dimension, KCT, covers the connection between knowledge about mathematics and about teaching. It includes knowledge of the advantages and disadvantages of certain representations and examples to introduce particular topics and to move learning forward, the ability to evaluate the difficulty of tasks and to develop instructional sequences. Lastly, knowledge of content and curriculum is derived from Shulman's category of curricular knowledge and was provisionally described as a category on its own. However, the research team left open the issue of it being part of KCT, running across various dimensions or constituting a separate domain (for more details, see Hill, Ball & Schilling, 2008; Ball et al., 2008).

A remarkable contribution of these projects is the measurement instruments they developed to assess teachers' mathematical knowledge for teaching (Hill et al., 2004). Although their elaboration is based in the US, and thus the multiple-choice items may be culturally influenced, various test forms were validated in other countries, including Chile (Martínez, Martínez, Ramírez & Varas, 2014). The tests usually include items for the four main dimensions of the model, i.e., CCK, SCK, KCS and KCT, but focuses mainly on the two categories related to subject matter knowledge (Hill et al., 2004).

Another important contribution of this research group is that, with the aid of these tests, they found that there is a significant relationship between teachers' specialized mathematical knowledge and primary school students' achievement gains in mathematics (Hill, Rowan & Ball, 2005). With this finding, they provided conclusive evidence about the relevance of both teachers' mathematical and specialized knowledge for teaching mathematics effectively.

However, the MKT model has also received criticism mainly for two reasons. First, the dimensions in the model are hard to differentiate. Theoretically, it is difficult to clearly discern what knowledge belongs to the dimension of specialized content knowledge and what to the dimensions of knowledge of content and teaching and knowledge of content and students (Döhrmann, Kaiser & Blömeke, 2018). Empirically, factor analyses conducted on teachers' answers to the MKT test do not support the distinction of all the categories included in the MKT model (Baumert et al., 2010; Hill et al., 2004). Despite this critique, "this test seems to provide a good overall assessment of mathematical knowledge for teaching" (Baumert et al., 2010, p. 141). A second area for criticism is related to its cognitive perspective. The model only includes knowledge facets, thus ignoring the specific context in which it is needed and teachers' beliefs and motivational aspects (Depaepe et al., 2013).

1.1.2.3 The COACTIV Study

A second large-scale project aiming at theoretically and empirically clarifying the professional knowledge needed by teachers is the German project Professional Competence of Teachers: Cognitively Activating Instruction, and the Development of Students' Mathematical Literacy (COACTIV). Its theoretical model of teachers' professional competence is grounded on approaches to teacher knowledge as the core of teachers' professional activity, in line with Shulman, integrating them with Weinert's view on professional competence. Baumert and Kunter (2013, p. 27) define the term competence as "the personal capacity to cope with specific situational demands" and emphasize that competence can be learned and, thus, taught. The integration of the competence concept in a broad sense, led them to include, besides specific knowledge, also motivational, metacognitive, and self-regulatory aspects as parts of professional practice.

The model of professional competence developed in COACTIV is a non-hierarchical model, for which the knowledge component was specified for the context of teaching. The model distinguishes "between four *aspects* of competence (knowledge, beliefs, motivation and self-regulation), each of which comprises more specific *domains* derived from the available research literature. These domains are further differentiated into *facets,* which are operationalized by concrete indicators" (Baumert & Kunter, 2013, p. 28). Within the professional knowledge aspects, two domains are subject-related, namely content knowledge and pedagogical content knowledge, and other three domains can be characterized as a broadened version of Shulman's general pedagogical knowledge, including general pedagogical and psychological knowledge, organizational knowledge and counseling knowledge. Content knowledge (CK) was conceptualized as "a profound mathematical understanding of the content of the curricular content to be taught" (Baumert et al., 2010, p. 142) that goes beyond school knowledge and everyday mathematical knowledge. Moreover, CK is assumed to be distinguishable from and a necessary condition for pedagogical content knowledge (PCK). PCK, in turn, is assumed to be "needed over and above CK to stimulate insightful learning" (p. 145). Baumert and Kunter (2013) characterized PCK as including three facets:

- Knowledge of the didactic and diagnostic potential of tasks, their cognitive demands and the prior knowledge they implicitly require, their effective orchestration in the classroom, and the long-term sequencing of learning content in the curriculum
- Knowledge of student cognitions (misconceptions, typical errors, strategies) and ways of assessing student knowledge and comprehension processes
- Knowledge of explanations and multiple representations. (p. 33)

The distinction between CK and PCK was empirically confirmed, also indicating a correlation between both factors, which contributes to supporting the "assumption that PCK as a specific form of mathematical knowledge is unconceivable without sufficient CK but that CK cannot substitute PCK" (Baumert et al., 2010, p. 166). Furthermore, their results indicate an empirical relation between teachers' knowledge and instructional quality and, thus, with student achievement. The effect of PCK on high-quality teaching and student learning was found to be empirically larger than that of CK, as PCK "largely determines the cognitive structure of mathematical learning opportunities" (p. 166). However, they argue, "this does not imply that CK [...] is unimportant. [...] CK defines the possible scope for the development of PCK and for the provision of instruction offering both cognitive activation and individual support" (p. 166–167).

1.1.2.4 The International TEDS-M Study

A third important large-scale study that has also theoretically and empirically investigated teachers' professional competencies is the Teacher Education and Development Study in Mathematics (TEDS-M) (Blömeke & Kaiser, 2014; Döhrmann, Kaiser & Blömeke, 2014; Tatto, Schwille, Senk, Ingvarson, Peck & Rowley, 2008). In contrast to the LMT and COACTIV projects, TEDS-M was a large-scale comparative international study and focused on preservice teachers rather than on in-service teachers. Preservice primary and secondary teachers in their last year of teacher education from 17 countries participated in the study. Its main purpose was to understand how primary and secondary teachers are prepared in the participating countries, looking at both national policies and institutional practices, and to investigate the relationship of these with preservice teachers' mathematical content knowledge and mathematical pedagogical content knowledge.

In order to measure the outcomes of teacher education, TEDS-M developed a framework to conceptualize teachers' mathematical competencies. Similarly to COACTIV, the TEDS-M framework is based on Weinert's (2001) conceptualization of competence. Hence, it includes cognitive facets that are understood as teachers' professional knowledge and beliefs and personal traits such as motivation and self-regulation. The measurement of multidimensional dispositions allowed the examination of the relationships between different facets and to build different teacher profiles (Blömeke & Delaney, 2012). Regarding the professional knowledge of teachers, the TEDS-M model was based on Shulman's (1986) approach to teachers' knowledge. Thus, it is divided into three facets: mathematics content knowledge (MCK), mathematics pedagogical content knowledge (MPCK) and general pedagogical knowledge (GPK).

Preservice teachers' MCK refers to the knowledge of the discipline, its structure and principles, understanding of mathematical concepts, ideas, facts and procedures (Tatto et al., 2008). Regarding its coverage, it is "expected to cover from a higher and reflective level at least the mathematical content of the grades the teacher would teach" (Döhrmann et al., 2014, p. 436). The items were categorized into three levels of difficulty, according to the curricular level of the content included. The *novice* level was given to items covering content typically taught at the school grades for which preservice teachers were being prepared to teach, the *intermediate* level indicated that the content was one or two grades above the highest grade included in the teacher education program, and *advanced* items were those covering content that is three or more years beyond the grades they were being prepared to teach. To organize the assessed subject matters, the content domains from the TIMSS study were used so that internationally dominating mathematics education topics were covered, namely number, geometry, algebra and data. These content domains, together with the three cognitive domains used in TIMSS—knowing, applying and reasoning—constituted organizing categories to build the test items.

The conceptualization of MPCK focused mainly on the core task of teaching, i.e., on tasks teachers do when planning and performing their mathematical lessons, including analyzing and evaluating student's work (Döhrmann et al., 2018). The domain of MPCK also made use of the three content domains and the three cognitive domains from TIMMS. Additionally, two sub-domains of MPCK were distinguished: curricular knowledge and knowledge of planning for mathematics teaching and learning on the one side, and knowledge for enacting mathematics for teaching and learning on the other side (Döhrmann et al., 2014; Tatto et al., 2008). The TEDS-M conceptual framework (Tatto et al., 2008, p. 39) describes the first sub-domain as a pre-active category, in which the knowledge needed for preparing mathematical lessons is included. For instance, to organize instructional activities, teachers need not only knowledge about the mathematics curriculum and the connections within it, they also need to be able to establish appropriate learning goals, identify key ideas in learning programs and know about different assessment formats. Additional knowledge and abilities required to concrete planning for mathematics teaching and learning include linking didactical methods and instructional designs, selecting and planning appropriate teaching strategies and learning activities. Teachers also have to select an adequate approach to representing mathematical ideas, anticipate different methods for solving mathematical problems, predict typical student answers and possible misconceptions and plan assessment options accordingly. The second sub-domain refers to the knowledge needed for enacting those planned lessons. Besides the ability to represent and provide appropriate explanations of mathematical ideas and generating productive questions that guide the learning process effectively, it covers the ability to analyze and diagnose students' thinking by, for example, evaluating and interpreting students' solutions and arguments, identifying and understanding their misconceptions and providing appropriate feedback.

Although TEDS-M included MPCK and MCK as separate domains in their framework and measured both domains in the study, their results indicated that it is very difficult differentiating completely between MCK and MPCK. Döhrmann et al. (2014) state that "it is impossible to construct disjoint sub-domains, because the solution of an item in the domain MPCK generally requires MCK" (p. 336). This finding is consistent with those of other studies (e.g., Krauss et al., 2008).

The third component of professional knowledge in the TEDS-M framework, i.e., general pedagogical knowledge (GPK), was not assessed as part of the main study. Only three countries opted to participate in a supplementary study that evaluated GPK using an instrument developed by König, Blömeke, Paine, Schmidt and Hsieh (2011). Despite not evaluating it, GPK was recognized in the TEDS-M framework as an essential component for effective mathematics teaching (Tatto et al., 2008) and it was characterized as including four knowledge categories: knowledge of students, knowledge of classroom management, knowledge of instructional design and knowledge of diagnostics and assessment. Results of the comparative study revealed differences between preservice middle school teachers from the three participating countries, particularly in the subdimension related to the generation of classroom management strategies (König et al., 2011). Additionally, examination of the results of preservice primary school teachers provided evidence for connecting GPK with MPCK (see König & Blömeke, 2010).

Besides knowledge facets, TEDS-M measured preservice teachers' beliefs, that are included in the affective-motivational facet of their theoretical model. They used a definition by Richardson (1996) that defines beliefs as "understandings, premises or propositions about the world that are felt to be true" (p. 103). The study differentiated between beliefs about the nature of mathematics, beliefs about the teaching and learning of mathematics, beliefs about mathematics achievement and beliefs about teacher education effectiveness and their own preparedness for teaching mathematics (Tatto et al., 2008). In particular, the scales on epistemological beliefs about the nature of mathematics differentiated between a static and formal and a dynamic and applied view of the subject (Blömeke & Kaiser, 2014). These beliefs are, in turn, assumed to influence teachers' beliefs about the teaching and learning of mathematics, which are also separated into transmissive and constructivist beliefs.

Kaiser et al. (2017) highlight the importance of the inclusion of beliefs facets besides the knowledge facets in the TEDS-M framework, as they add to the model the possibility to relate achievement profiles to the beliefs of preservice teachers. In fact, Blömeke, Suhl, Kaiser and Döhrmann (2012) found that two achievement profiles could be identified in most participating countries. On the one side, preservice teachers with strong MCK and MPCK and constructivist beliefs and, on the other side, preservice teachers showing low results in knowledge tests and more transmission-oriented beliefs. Moreover, König (2012) suggests that beliefs have the role of connecting teachers' knowledge and performance, mediating their decisions, especially in challenging situations. Blömeke (2012) indicates that it is feasible to assume that "a certain level of knowledge is needed before it is possible to see the dynamic nature of mathematics" (p. 18) and that beliefs about mathematics learning are also influenced by epistemological beliefs. Thus, teachers who are able to see mathematics as a dynamic subject could also recognize the benefits of student-centered and explorative teaching strategies. Actually, in a study comparing the beliefs of preservice secondary teachers from four countries, Blömeke (2012) found that "higher MCK and MPCK increases dynamic beliefs on the nature of mathematics and decreases static beliefs [...]. These beliefs, in turn, influence how a teacher regards teaching and learning, either from a more constructivist or from a transmission point of view" (p. 32). Thus, higher professional knowledge would lead, indirectly, to increased student-oriented teaching practices.

In summary, the TEDS-M study measured both cognitive and affective-motivational characteristics of preservice primary and secondary teachers in 17 countries. Preservice teachers' MCK and MPCK and their beliefs were evaluated in every participating country, whereas GPK was assessed only in Germany, Taiwan and the USA. Results from the study provided cross-national evidence about the characteristics of preservice teachers' competence for teaching mathematics and its disparities both across and within countries.

1.1.3   Advances in the Conceptualization of Mathematics Teachers' Professional Competence

Based on the current discussion on mathematics teachers' professional competences, Kaiser et al. (2017) differentiate between two approaches. The first approach focuses on the cognitive aspects of (preservice) teachers' professional competence, specifically on the knowledge facets needed by teachers to effectively do their work. This cognitive perspective has been dominating in recent decades and includes studies such as LMT, COACTIV and TEDS-M. As has been described above, these studies take an analytical approach in that each of them distinguishes between a number of knowledge facets and, some of them, also affective-motivational facets that together form the structure of teachers' professional competence. Thus, teaching is modeled in a multi-dimensional way, in which Shulman's (1986) categorization of teachers' knowledge can be recognized at the core. A second approach includes newer studies that have brought to the foreground context and situated aspects of the teaching and learning process, departing from the concept of noticing (Kaiser et al., 2017). Kaiser and König (2019) point out that "these studies assume the multidimensionality of referring not only to subject-based cognitive aspects but also to pedagogical reflections on the teaching-and-learning as a whole" (p. 600).

Depaepe et al. (2013) also recognized these perspective differences in relation to the study of pedagogical content knowledge and argued that they impact the way teachers' knowledge is empirically studied. Studies under the cognitive perspective measure knowledge "independently from the classroom context in which it is used, most often through a test" (p. 22), whereas studies taking a situated perspective consider that knowledge only makes sense in the context it is used and, thus, should be measured by observing it in the classroom.

After unpacking these apparently opposed approaches, Blömeke, Gustafsson and Shavelson (2015) proposed a framework that conceptualizes teachers' competence as a continuum integrating both cognitive traits and situational aspects. They argue that dichotomizing both approaches is misleading and unproductive because both perspectives have pitfalls and fail to cover the complexity of the construct, either by ignoring the relevant dispositions that need to be in place or by neglecting the observable and context-specific behavior. By systematically outlining conceptual and assessment controversies between both approaches, they identified commonplaces and ways in which they can complement each other. In their proposed model, competence is regarded as a continuous process, in which certain disposition traits have to be in place so situation-specific skills can be activated to generate the observable performance of a teacher in the context of their classroom. Disposition traits include cognitive aspects, such as teachers' professional knowledge, and affective-motivational aspects such as beliefs. Situation-specific skills aid the process of connecting the disposition facets and integrating them into observable performance by activating processes of perception, interpretation and decision-making. Furthermore, they suggest that their model may serve as a tool for research on competence development and that "the measurement of competence, then, may be viewed along a continuum from traits (cognitive, affective, motivational) that underlie the perception, interpretation, and decision-making that give rise to observed behavior in a particular real-world situation" (p. 11). Although applying such a framework for research is conceptually and methodologically challenging, they point out that it would be a valuable tool in overcoming the risk of neglecting either the cognitive dispositions or the observable behavior.

1.1.3.1 Expertise and Noticing

Another perspective on studying teachers' competences has focused on their expertise. Expertise research aims at identifying characteristics, skills and abilities of experts under the assumption that "examining and knowing the nature of expertise also helps us understand what it may take for a novice to become an expert in that field" (Li & Kaiser, 2011, p. 3). In particular, understanding the nature of expertise in complex fields involving multiple and varied tasks, such as in the teaching profession, presents important challenges. Besides the lack of clarity about what constitutes effective practice, there is a wide variety of factors that play a role in the task of teaching, which takes place in a particular cultural context that also needs to be accounted for (Li & Kaiser, 2011). A common approach for studying mathematics teachers' expertise involves examining the knowledge and skills expert teachers' put in place when teaching mathematics. However, identifying and selecting expert teachers has also proven to be a challenging task (Berliner, 2001). Yet another approach to understanding teachers' expertise is to compare experts' and novices' performances. This allows identifying the aspects that significantly differentiate an expert teacher from a novice teacher and also to describe the development of expertise in teaching. Some identified differences have pointed to the flexibility of expert teachers, as they are more able to adapt their teaching in response to students (Berliner, 2001). Chi (2011) states that a relevant difference is on the perspective under which experts organize knowledge, that allows them to have and make use of a deeper and more connected and structured representation of topics. Expertise research has also focused on comparing novice and expert teachers' situation-specific skills, on how they perceive and interpret classroom incidents. In particular, the *noticing* framework has become prominent in educational research, with a large influence from the work of van Es and Sherin (2002).

Teachers' ability to notice has been recognized as a component of teachers' expertise that is especially relevant for current views of teaching and learning, in which teachers are asked to adapt their instruction based on students' thinking (Sherin, Jacobs & Philipp, 2011; van Es & Sherin, 2002). For teachers to make pedagogical decisions during a lesson according to students' mathematical ideas, first, they have to be able to attend to and interpret relevant students' ideas and other key interactions in the midst of the complex and constantly changing environment of the classroom. To notice can therefore be described as teachers' ability "to attend to aspects of classroom interactions that influence student learning and reason about them in the midst of instruction" (van Es, 2011, p. 134). Furthermore, "learning to notice in particular ways is part of the development of expertise" (Jacobs et al., 2010, p. 170). In fact, in a review of research in the field, Stahnke, Schueler and Roesken-Winter (2016) found that several studies have concluded that expert teachers tend to show higher noticing skills and to notice more relevant classroom situations.

In their overview of the construct of teacher noticing, Sherin, Jacobs & Philipp (2011, p. 5) describe it as an active process, in which teachers do not only observe but also take part in the classroom situation of which they are trying to make sense and to manage. They state that, although diverse conceptualizations are used by different researchers, two aspects can be recognized as a commonality: attending to and making sense of particular events in the teaching context. They explain that "to manage the complexity of the classroom, teachers [...] must choose where to focus their attention and for how long and where their attention is *not* needed and, again, for how long". They continue to explain that, after attending to specific features or moments in the classroom, teachers "interpret what they see, relating observed events to abstract categories and characterizing what they see in terms of familiar instructional episodes." Furthermore, they emphasize that these two aspects are not totally distinct in that what teachers attend to is related to what they are intending to make sense of, their interpretations also affect what they pay further attention to and their decisions also produce new teaching and learning situations to which they potentially could attend to.

Despite these common aspects of the construct, a variety of conceptualizations exist among researchers, with different emphases. For instance, Star, Lynch and Perova (2011) investigate what teachers attend to and what they filter out and do not attend to. They suggest that identifying significant classroom features is the most fundamental aspect of noticing. They argue that "if preservice teachers are unable even to identify that classroom events have occurred [...], it seems natural that they will be unable either to make connections between these events and broad principles of teaching and learning [...] or to reason about these events" (pp. 119–120). They claim that preservice teachers must first activate their ability to notice all kinds of classroom features and only after they have developed this ability they can attempt the more sophisticated ability of differentiating between important and trivial events. In a study also focusing only on teachers' attention to events by wearing cameras during a lesson, selecting important events on the moment and then reflecting on those events, Sherin, Russ and Colestock (2011) found that "teachers often selected moments to capture because of how well they aligned with their expectations for their lessons" (p. 90). Sometimes teachers selected moments because they differed from what they expected and they found them surprising, and in other cases because they confirmed what they were thinking it would occur.

In addition to the attending facet of noticing, van Es and Sherin (2002; 2006) emphasize the interpretation facet. They state that "how individuals analyze what they notice is as important as what they notice" (van Es & Sherin, 2002, p. 575). They propose that after identifying what is noteworthy in a particular instructional situation, teachers use their knowledge of the context, of the particular students and their professional knowledge to make connections and reason about what they decided to attend to. Reasoning about certain features of an instruction situation and connecting it with broader issues of teaching and learning are crucial for teachers to make sense of the events in a classroom. In the process of trying to make sense of students' thinking, teachers use the evidence they observe in the classroom and recognize it as a part of broader teaching and learning principles (van Es, 2011). Consistent with this view of noticing and with the need to make connections to make productive interpretations, Goldsmith and Seago (2011) add that interpreting noteworthy events "involves attending to both the mathematical content of the task and students' mathematical thinking" (p. 170). In a study that promoted the development of preservice teachers' interpretation skills by engaging them with a software that guided them in interpreting classroom interactions based on evidence, van Es and Sherin (2002) found that, in comparison to a control group, preservice teachers who used the software became more analytical and increased the use of specific evidence to support their interpretations about students thinking and mathematics teaching and learning.

Another conceptualization of teacher noticing includes, besides attending to relevant events and interpreting them, the feature of deciding how to respond. In their framework, Jacobs et al. (2010) decided to focus on teachers' noticing of children's mathematical thinking and conceptualize this construct as "a set of three interrelated skills: attending to children's strategies, interpreting children's understandings, and deciding how to respond on the basis of children's understandings" (p. 172). Their research looks particularly at the extent to which teachers attend to details in the mathematical strategies used by students, at how they interpret students' mathematical understandings by using evidence on the details of the strategies used by students and connecting it with research on children's development and mathematical learning. In contrast to other research groups, they also focus on the way teachers articulate their pedagogical decisions based on what they have interpreted about students' mathematical thinking. In other words, they also investigate if teachers' instructional decisions are consistent with their interpretations of students' strategies. They argue that:

> what one notices cannot be separated from one's goal for noticing. [...] a teacher's goal in attending to and making sense of children's thinking is to use that information to make effective instructional responses. Because of the nature of teachers' work, these responses must generally be made quickly, often, and in the midst of instruction. Thus we argue that responding should be considered a component skill of professional noticing of children's mathematical thinking because identifying, describing, interpreting, and responding are all interrelated. (Jacobs, Lamb, Philipp, Schappelle & Burke, 2007, p. 8)

In a subsequent study (Jacobs, Lamb, Philipp & Schappelle, 2011), they further highlight that the use of these three skills is closely connected and occur almost simultaneously during teaching and should therefore be kept together in the noticing framework so that the goal of making effective and responsive decisions remains visible. They used a cross-sectional design to study the development of expertise in professional noticing of children's mathematical thinking and concluded that "teachers need support in learning to attend to children's strategies, and they need additional support to learn how to use those details in deciding how to respond so that their instruction maintains children's thinking as central" (p. 111).

A similar conception of teacher noticing is found in the work of Santagata and colleagues on the Lesson Analysis Framework (Santagata, 2011; Santagata & Yeh, 2016; Santagata, Zannoni & Stigler, 2007). They have focused on teachers' abilities to analyze classroom lessons. These analyses require abilities that are similar to the skills in teacher noticing. Teachers are encouraged to selectively attend to classroom events and use their knowledge to reason about noteworthy moments of the teaching and learning process, focusing especially on the interpretation of students' thinking and learning. In addition, similar to Jacob and colleagues' framework, they extend their framework to a third process, in which teachers generate alternative strategies that would improve the effectiveness of the lesson by enhancing students' opportunities to learn mathematics.

The work of Jacobs et al. (2011) is particularly relevant to the situation-specific skills described by Blömeke et al. (2015). There is an evident parallelism between the three intertwined processes of noticing described by Jacobs and colleagues: *attending* to children's strategies, *interpreting* children's understandings and *deciding how to respond* on the basis of children's understandings and the three situation-specific skills placed at the center of Blömeke and colleagues' framework: perception, interpretation and decision making. However, there are also some differences between both models. One distinction is that the noticing model looks only at the three named processes, whereas the competence-as-a-continuum model relates the situation-specific skills to dispositions and observable behavior. Another difference is related to their focus. While Jacobs et al.'s (2011) noticing framework focuses on students' mathematics thinking, the competence framework of Blömeke et al. (2015) provides a wider competence framework that can model, among other competences, a broad range of aspects related to mathematics teaching.

1.1.3.2 TEDS-M Follow-up Studies

A prominent empirical study combining both cognitive and situated perspectives is the TEDS-FU study. As a follow-up to the TEDS-M study, it explored "how mathematics teachers' professional knowledge develops after the end of teacher education in the first years of their school career" (Kaiser, Blömeke, Busse, Döhrmann & König, 2014, p. 45). TEDS-FU builds on TEDS-M rather cognitive oriented framework and extends it to a situated approach. For doing this, it considers the expertise approach and the noticing framework as they highlight the importance of the integration of different dimensions of teachers' professional knowledge and emphasize an orientation towards teachers' performance in a contextualized way (Kaiser et al., 2017). From expertise research, the TEDS-M framework considers contributions stating that

> expert teachers perceive classroom situations more quickly, more accurately and more holistically than novice teachers. They anticipate more appropriately and faster what might happen in the teaching situation [...] expertise is characterized by a high degree of integration of knowledge with multiple links, a modified categorical perception of teaching situations and by increasing integration of the different dimensions of professional knowledge. (Kaiser, Busse, Hoth, König & Blömeke, 2015, p. 373)

Linked to the noticing framework and drawing on the construct of competence as a continuum proposed by Blömeke et al. (2015), the TEDS-FU framework elaborated on three situation-specific skills for mathematics teachers in the PID-model: (a) *Perceiving* particular events in an instructional setting, (b) *Interpreting* the perceived activities in the classroom and (c) *Decision-making,* either as anticipating a response to students' activities or as proposing alternative instructional strategies (Kaiser et al., 2015, p. 374). This conceptualization of situation-specific skills applies to a broad range of instructional aspects that are important for effective and quality mathematics teaching, in contrast to the specific focus on students' mathematical thinking of some approaches on the noticing framework (Jacobs et al., 2011).

Empirical examination of such a broad conceptualization of mathematics teachers' professional competence requires a range of different assessment instruments. The TEDS-FU study used web-based tests (Kaiser et al., 2017). To evaluate cognitive disposition facets, digitalized and shortened versions of the TEDS-M tests for MCK, MPCK, and GPK were created. To cover the situated PID facets, video-based assessments were developed. The items were designed to evaluate both demands of teaching related to mathematics (M_PID) and demands related to pedagogy (P_PID). This exemplifies how the PID-model can be applied to a broad range of instructional issues. The items comprised different areas of mathematics (number, operations, patterns and structure and geometry), a variety of mathematics competences such as modeling, problem-solving and arguing, and different lesson phases. Regarding pedagogical demands, items included issues of classroom management, heterogeneity in the classroom, individualization of teaching and teaching strategies (Blömeke et al., 2015; Kaiser et al., 2017). Items embraced the three parts of the noticing process and were designed to evaluate low-inference and high-inference levels of noticing expertise. In addition, a time-limited test in which preservice teachers were prompted to identify typical student errors was included (Pankow, Kaiser, Busse, König, Blömeke, Hoth & Döhrmann, 2016) to evaluate teachers' diagnostic competence.

The study collected follow-up data from primary and secondary teachers who also participated in the TEDS-M study. Analyzing the data from secondary mathematics teachers, Kaiser and colleagues (2017) found strong correlations between the measures of noticing related to mathematics (M-PID) and those related to pedagogy (P_PID), confirming that the items evaluate connected constructs that are included under the overarching concept of noticing. At the same time, they found low correlations between both mathematics components (MPCK and M_PID) and between both pedagogical components (GPK and P_PID). Thus, providing evidence for the distinction between the cognitive and situated perspectives and

illustrating how both approaches can be complemented to reach a more comprehensive understanding of teachers' competence and its development. Furthermore,
the researchers point out that taking a cognitive or a situated approach as an
underlying paradigm for empirical research affects the selection or design of
assessment instruments and the results obtained and conclude that "both ways
of testing capture specific characteristics of teachers' professional competencies
and can therefore be regarded as necessary to display their richness and variety"
(p. 178).

1.2   Diagnostic Competence

Among the wide range of demands faced by teachers in their daily instructional
activities, the adaptation of pedagogical strategies to students' needs, knowledge and abilities is crucial for good quality lessons (Darling-Hammond, 2000;
Helmke, 2017; KMK, 2014; NCTM, 2000). Heterogeneous classrooms require
teachers to differentiate and individualize strategies and tasks amongst learners,
who most often show a variety of learning speeds, styles and needs. Therefore,
teachers have to gather information during the students' learning process that
informs them about the understanding of individual students or groups of students. The interpretation of information gained through during-lesson observation,
analysis of student work, making questions to students or formal testing is then
used to make in-the-moment decisions, to plan further teaching or to allocate specific support or resources. Teachers' knowledge, abilities, motivation and beliefs
required to go through this process aimed at analyzing and understanding student
thinking have been conceptualized as *diagnostic competence* (Leuders, Dörfler,
Leuders & Philipp, 2018; Prediger, 2010; Schrader, 2013).

Despite its relevance for effective teaching, the construct of diagnostic competence has been used in reference to diverse aspects of teachers' activities, revealing
not only distinct research traditions and cultures but also differing conceptualizations and models (Barth & Henninger, 2012; Leuders et al., 2018; Prediger, 2010).
The following sections present and describe some of these conceptualizations and
models.

1.2.1   Diagnostic Competence as Accuracy of Judgments

One branch of studies on teachers' diagnostic competence has focused on the
accuracy of teachers' judgments about student traits (Helmke & Schrader, 1987;

Hoge & Coladarci, 1989; Praetorius, Berner, Zeinz, Scheunpflug & Dresel, 2013; Schrader, 2006; Südkamp, Kaiser & Möller, 2012). Diagnostic competence is conceptualized as the agreement between teachers' judgments and some objective measure of student achievement or observed responses to a questionnaire. Helmke and Schrader (1987) operationalized it as "the correlation between a teacher's predicted scores for his or her students and those students' actual scores" (p. 94). The method varies among studies according to characteristics of the data, but in essence consists of evaluating to what extent the average, variance or rank order of teachers' prediction is consistent with actual data obtained from the students. Teachers' judgments of students' performance are collected and, at the same time, students are tested on the same particular trait that can be either a cognitive or a non-cognitive trait. Both values are then compared, and the level of agreement is taken as a measure of teachers' diagnostic competence (Helmke & Schrader, 1987; Hoge & Coladarci, 1989; Schrader, 2013; Spinath, 2005).

The relevance of judgment accuracy is grounded on the impact teachers' judgments have on various teaching decisions, especially on those related to adapting instructional activities to account for differences between learners and individual students' needs (Helmke & Schrader, 1987; Hoge & Coladarci, 1989; Praetorius et al., 2013). In other words, teachers' "decisions based on accurate assessments of student attributes will be more functional than those based on inaccurate assessments" (Hoge & Coladarci, 1989, p. 308). In fact, in an empirical study, Helmke and Schrader (1987) asked teachers to predict students' performance on a mathematics test evaluating the contents they had covered in recent lessons. They found that high judgment accuracy in combination with frequent use of structuring cues and individualized supportive contact substantially affected achievement gains. Their results indicate that high cognitive growth occurred when teachers' strong diagnostic competence was combined with a high-frequency use of structuring cues. Moreover, the lowest learning growth was related to teachers' strong diagnostic competence combined with low instructional quality, showed by a low frequency of structuring cues and individual support. Helmke and Schrader (1987) interpret this effect stating that "if teachers either do not possess effective teaching skills or if they are – for whatever reasons – not motivated to make use of them, high diagnostic sensitivity appears to have a detrimental effect on students' course of achievement" (p. 96). In addition to consequences for instructional decisions, Südkamp et al. (2012) suggest that teachers' judgments also affect the early identification of students with difficulties in school, their further evaluation and provision, and can influence teachers' expectations about students' ability, having consequences for students' academic careers and self-concepts.

In order to more precisely define what is meant by accuracy, Schrader and Helmke (1987) introduced the concept of veridicality and, based on Cronbach (1955), outlined three components for assessing accuracy. Veridicality describes the accuracy of teachers' judgment in terms of the agreement between the objectively measured student trait and the evaluation or prediction of that same trait by the teacher. Veridicality is assessed by examining three components. The first is the level component, which considers teachers' tendency either to under- or overestimate the performance level of the whole class. The second constituent is the differentiation component that characterizes teachers' propensity to either over- or underestimate the variance among students' performance in a class. The third element is the comparison component and refers to the evaluation of the relative performance of the students in relation to one another, for example, in the form of a ranking. The comparison component, also named rank component, is described by Schrader and Helmke (1987, p. 33) as diagnostic sensitivity in the narrower sense since it reveals the ability of the teacher to diagnose or rank the characteristics of the students in their class independently of global judgment tendencies.

Hoge and Coladarci (1989), in their meta-analysis, point out that the degree of specificity of teachers' judgments varies among studies according to the types of judgment measures employed:

> (a) *ratings* (low specificity), where teachers rated each student's academic ability (e.g., "lowest fifth of class" to "highest fifth of class"); (b) *rankings*, where teachers were asked to rank order their students according to academic ability; (c) *grade equivalence*, where teachers estimated, in the grade-equivalent metric, each student's likely performance on a concurrently administered achievement test; (d) *number correct*, where teachers were asked to estimate, for each student, the number of correct responses on an achievement test, administered concurrently; and (e) *item responses* (high specificity), where teachers indicated, for each item on an achievement test administered concurrently to the students, whether they thought the student would respond correctly to the item or had sufficient instruction to respond correctly. (Hoge & Coladarci, 1989, p. 300 f)

The meta-analyses of Hoge and Coladarci (1989) and Südkamp et al. (2012) reviewed studies examining the correspondence between teachers' judgments of students' performance and students' observed achievement. Hoge and Coladarci's (1989) examined 16 studies and concluded that teachers' accuracy of judgments is, in general, valid and large. Moreover, their results indicate that direct teachers' judgments (when the criterion to which teachers' judgments were compared was made explicit to teachers) correlate stronger with observed achievement than indirect judgments (when teachers are asked to estimate general achievement rather than performance on a test). The more recent meta-analysis by Südkamp et al. (2012) examined 75 studies on diagnostic accuracy and concluded that the overall mean effect size was .63 and significant. However, they also found a large variance on effect sizes across studies, suggesting the need to include moderator variables in the model. Their analyses identified two relevant moderators. First, their results show that teachers' informed judgments (studies in which teachers were informed about the test to which their judgments would be related) are related to higher judgment accuracy than teachers' uninformed judgments. The second significant predictor was the relation between judgment and test characteristics. Teachers' judgments were found to be more accurate when there was congruence of the evaluated domain and abilities between their judging task and the achievement test administered to students.

As part of a study about the accuracy of teacher judgments on student achievement, Spinath (2005) investigated the relationship between the three components of accuracy suggested by Schrader and Helmke (1987). In doing so, she examined the correlations between the different accuracy components for the same student trait and the correlations between the same accuracy component across different student traits. Her results indicate that there are no significant correlations, suggesting that accurate teacher judgments cannot be considered to be at the base of a general diagnostic competence. Moreover, she advocates that statements about the accuracy of teacher judgments should make explicit the types of student characteristics and the accuracy component they refer to. Thus, she concludes that "the term diagnostic competence should not be used to refer to an ability to judge pupils' attributes accurately" (Spinath, 2005, p. 85).

In line with the need to specify the accuracy component and the student characteristic being examined, the results of COACTIV also suggest that diagnostic competence should be viewed as a multi-dimensional ability construct (Brunner, Anders, Hachfeld & Krauss, 2011). Based on the operationalization of diagnostic skills as the correlation between teachers' judgments and students' observed achievement, the COACTIV study investigated various indicators of teachers diagnostic skills, such as teachers' judgment on whole class' achievement level and distribution, motivational level, the percentage of high-achieving and low-achieving students in their class and individual students' answers to specific tasks. Teachers were also asked to rank the order of achievement of seven students and to evaluate the demand of four tasks by determining the percentage of students that would solve them correctly. Their results show very small correlations among the various indicators of diagnostic skills of secondary mathematics teachers. Thus, no one-dimensional model could be found, suggesting that it should be considered a multi-dimensional construct (Binder, Krauss, Hilbert, Brunner, Anders & Kunter, 2018; Brunner et al., 2011).

In fact, these results are one of the reasons the term *diagnostic skills* is used in the COACTIV study instead of the term *diagnostic competence*. Additionally, Brunner et al. (2011) argue that using the concept of skills is helpful in differentiating the construct from the much wider construct of teachers' professional competences and for highlighting that the examined skills are only some key indicators from the wider and multi-dimensional concept of *diagnostic expertise* as explained by Helmke (2003).

Helmke (2017; 2003) uses the concept of *diagnostic expertise* to emphasize the differences with the concept of *diagnostic competence* understood as the accuracy of judgments. He states that teachers perform various diagnostic activities, including both continuous implicit diagnoses of the constantly changing learning needs of students and explicit forms of collecting and processing information about students' learning that do not immediately inform the teaching process (Helmke, 2017, p. 121). Thus, he implies that diagnostic activities of teachers cover a wider range of demands than the mere accurate judgment of students' performance. Because the construct of diagnostic expertise is very broad conceptualized and considering that the present study only contemplates some aspects of it, the concept of diagnostic competence will be used and further specified.

1.2.2   Diagnostic Competence as Basis for Adaptive Pedagogical Decisions

Besides the conceptualization of diagnostic competence as teachers' judgment accuracy, many studies have taken a broader perspective in which teachers' diagnostic activities are not relegated to formal assessments, tests and grading, but are extended to diagnostic activities that occur both at lesson planning and during the teaching and learning process. Schrader (2013) describes this perspective shift stating that whereas previous approaches focused on the accuracy of teachers' evaluation of students' performance, the focus is now on teachers' collection and usage of information about students' performance, needs and learning processes that provide the basis for making pedagogical decisions. Also within this approach, a variety of attempts are being made to develop a wider theoretical model for describing diagnostic competence, which has led to diverse definitions and models for the same construct (Praetorius, Lipowsky & Karst, 2012).

Praetorius et al. (2012) point out that the conceptualization of diagnostic competence as accuracy of judgment is too distant from the concept of competence because the way teachers' judgments are collected is isolated from the real context and excludes any interaction with students. They argue that teachers' in-context appreciations of aspects directly related to students' learning processes are more important during instruction than judgments about students' performance on a test. Moreover, in-context assessments are cognitively more demanding because they take place under complex circumstances and under the pressure to make in-the-moment decisions that promote learning. Teachers' assessment of students' learning occurs in many situations during a lesson. For instance, teachers have to recognize the actual learning stage of individual students to be able to give appropriate feedback or support, they have to ask productive questions to understand students' thinking and they have to be able to anticipate students' learning paths and possible difficulties that may arise. Furthermore, Praetorius et al. (2012) recognize that for this broader conceptualization of diagnostic competence, teachers must also bring together pedagogical content knowledge and general pedagogical knowledge. This would raise some problematic issues for researchers related to the measurement of the competence, since it may be strongly correlated with other features of teachers' competences that may be difficult to differentiate empirically.

Similarly, Abs (2007) raises the issue that in their professional activities, teachers do not only have to deal with matters that need to be correctly measured. The demands of teaching include interacting with learners who receive teachers' diagnoses as feedback and have to analyze their results to learn further. Thus, he suggests that even though the accuracy of teachers' judgments may be an important part of diagnostic activities, it should not be considered as the single or central component of diagnostic competence. As indicated at the end of the previous section, with the aim of highlighting the differences with the accuracy of judgment approach, Helmke (2017) introduced the concept of diagnostic expertise, which he defines as a more wide-ranging construct that includes all the knowledge base related to pedagogical diagnostics. He described three components of diagnostic expertise, namely methodological and procedural knowledge, as well as conceptual knowledge, i.e., knowledge about methods for judging or predicting student achievement and knowledge about judgment tendencies and errors (Helmke, 2017, p. 119).

A different characterization of the components of teachers' diagnostic competence is provided by Prediger (2010). She described it as "a teacher's competence to analyze and understand student thinking and learning processes without immediately grading them" (p. 76). She emphasized the connection between teachers' ability to analyze and comprehend students' thinking and a student-centered teaching style, in which student thinking constitutes the starting point for further learning. Her characterization of the competence includes four components. First, teachers' interest in student thinking is at the base of the competence. Being curious about how students think mathematically is a requisite to further investigate their thinking processes. Yet, focusing only on students' reasoning deficits is shown to be insufficient. Second, an interpretative attitude is needed to understand student thinking from an inner perspective. In other words, adopting an analytic approach that instead of judging students' works aims at understanding the underlying reasoning of student thinking is a valuable component for diagnostic competence. Third, in a similar way as emphasized by Helmke (2003), theoretical knowledge about learning processes is needed as a fundament for analyzing student thinking. A theoretical framework that can guide diagnostic processes should be composed not only of general theories of the learning of mathematics but also by specific knowledge of the particular topic and its learning. Thus, the fourth component is domain-specific mathematical knowledge for teaching. This knowledge is specific for each topic of mathematics and includes different meanings and typical mental images for a concept that may help in unpacking students' thinking.

Similarly, von Aufschnaiter et al. (2015) specify three facets of teachers' diagnostic competence. They suggest a (subject-specific) diagnostic facet that refers to teachers' ability to select suitable diagnostic tools according to their goal and the occasion. Teachers have to be familiar with basic assessment methods so that they can choose from existing tools or develop their own ones that meet some reasonable quality criteria. In addition, diagnostic competence also includes other skills, such as the ability to make accurate judgments or considering a suitable standard as a reference. This first facet must be supported by specialized knowledge. Therefore, the two remaining facets refer to teachers' knowledge of findings and theories about learners' cognitive development on the one hand and their motivational and emotional characteristics on the other. This includes knowledge that can be used as criteria for the diagnosis and for the interpretation of the results by contrasting them with what should be typically expected.

Another approach to characterizing the construct of diagnostic competence has been to represent it as a process model. As with other approaches, different studies have developed different models with diverse emphases.

One series of studies taking such an approach is research on teachers' pedagogical thoughts, judgments and decisions. It has focused on the examination of the connections between teachers' intentions and their behavior, assuming that teachers make rational decisions when acting in their complex environments (Sha-velson & Stern, 1981). However, due to the need for immediate response in some or most teaching situations and the human tendency to constructing simplified models of very complex situations, this rationality refers more to teachers' judg-ments and intentions than to their real behavior. A cognitive model that identifies some factors influencing teachers' decisions was developed by Shavelson et al. (Shavelson, 1983; Shavelson, Atwood & Borko, 1977; Shavelson & Stern, 1981). The model considers that because the amount of information teachers have about students is very large, they select pieces of information and integrate them to make a judgment. Pedagogical decisions about content selection, tutoring or handling behavior problems are then made based on teachers' judgments about students' cognitive, affective and behavioral states (Shavelson et al., 1977; Shavelson & Stern, 1981).

Within this line of research, particular attention has been given to teacher's thoughts, judgments and decisions during interactive teaching. Synthetizing pre-vious research, Shavelson and Stern (1981) developed a model of teachers' decision making during interactive teaching. It considers that teachers usually build mental scripts or images of their plan for a lesson and try to maintain the flow of the activity. They constantly monitor the situation for determining if it is proceeding as planned. Decision making arises when student behavior exceeds a threshold of deviation from what was expected. First, teachers have to decide if immediate action is required and to consider if an appropriate solution is availa-ble and possible to apply to the situation (Peterson & Clark, 1978; Snow, 1972). Usually, teachers only consider alternative strategies when teaching is unsatis-factory and actually change strategies just in some situations (Peterson & Clark, 1978). However, the model does not specify the nature and characteristics of these decisions, nor does it indicate whether they are based on a diagnosis of student learning.

From a different perspective, Jäger (2006) proposed a model of diagnostic pro-cesses, which he described as the strategies leading to answering questions that are generally formulated at the beginning and specified later to develop either a diagnosis or a prognosis. The process occurs methodically and systematically, starting with a question and ending with an answer that is used for making deci-sions. In between, Jäger (1999, 2006) describes the process in five partial steps. The starting point is the formulation of a question, which reveals a lack of infor-mation that needs to be investigated in the form of a hypothesis. In order to run the process meaningfully and effectively, the question must be formulated precisely, which may include reformulating it. In a second phase, the question is expressed as a hypothesis that will be tested during the process. The third step consists of selecting methods and collecting data to study the hypothesis. The characteristics of tests or other assessment instruments need to be weighted in relation to the goals being pursued and their role in the diagnostic process. During data collection, new questions and hypotheses may arise. Based on the collected information, hypotheses are verified or rejected, and a diagnostic judgment is developed. Finally, the obtained diagnostic is reported and is often related to further measures or provisions for students. On the whole, it is evident that this diagnostic process refers to the use of objective measures and formal diagnostics and is, thus, not so closely related to the diagnoses teachers' have to make daily on an ongoing basis during each lesson.

Focusing also on the processes that lead teachers to their judgments, Wildgans-Lang, Scheurer, Obersteiner, Fischer and Reiss (2020) consider diagnosis as a process in which teachers draw inferences based on the assessment and interpretation of information. They adapted a model by Fischer et al. (2014), which distinguished eight epistemic activities for scientific reasoning and argumentation, and formulated it in terms of diagnostic activities. The eight activities are: problem identification, questioning or formulating the problem as a question that may guide the next steps in the process, generating hypotheses about possible sources of the error or learning difficulty, construction or selection of useful artifacts for identifying the student's understanding drawbacks, generation of evidence by using the evaluation artifacts with the student, evaluating the evidence obtained from the student's solution or answers, drawing conclusions and finally communicating the results to students, parents or other professionals. The model contemplates that teachers can engage in any of the activities multiple times and modify the order of the process according to the needs of the situation.

With a more specific focus on the skills that teachers need in diagnostic situations during instruction, Edelenbos and Kubanek-German (2004) referred to diagnostic competence in language education as "the ability to interpret students' foreign language growth, to skillfully deal with assessment material and to provide students with appropriate help in response to this diagnosis" (p. 260). It is evident that this definition does not base the quality of diagnostic competence on the accuracy of judgment but rather emphasizes the skills and abilities that teachers require for diagnosing, evaluating, interpreting and supporting the learning process. In fact, they acknowledge that a certain level of precision is required in any assessment, but in their operationalization of the competence, they state that it is composed of teachers' pedagogical attitudes towards learners, abilities such as perceiving, interpreting and analyzing, assessment skills and capacities to scaffold instruction according to the conclusions of the diagnosis. Edelenbos and

Kubanek-German (2004) identified six levels of primary school teachers' diagnostic competence that can be observed in various classroom activities. These levels start with intuitive and general observations and evolve towards an ability to bring together several observations, and to select, adapt and interpret assessment instruments. Teachers in the highest level are able to develop rich interpretations of learning situations, which become increasingly complex as they are made during the ongoing processes of teacher-student interactions.

Similarly, Barth and Henninger (2012) argue that "a competent diagnosis in teaching situations means that a teacher has to be able to recognize how students indicate their learning requirements within the scope of social interaction" (p. 50). Moreover, they suggest that competent diagnosis in classroom situations is composed of five dimensions, namely planning lessons in a way that allows for diagnosis of students' learning needs, perceiving the situation, generating hypotheses and testing them, recognizing and reflecting on own's subjectivity on interpretations and showing receptive and productive communicative behavior.

Also from a classroom situation approach and based on Jäger's (2006) conceptualization of diagnosis as a process, Klug, Bruder, Kelava, Spiel and Schmitz (2013) developed a cyclical process. Rather than focusing on the accuracy of teachers' judgments of students' achievement, the model focuses on teachers' diagnoses of students' learning behavior. They define learning behavior as "observable behavioral patterns that children display as they approach and undertake school learning tasks" (p. 39). Teachers' understanding of learning behavior should enable them to adapt instruction to their students' needs and to support students individually. The model has three dimensions, taking place in a preactional, an actional and a postactional phase. In the preactional phase, teachers decide on their diagnostic goals, which implies an explicit intention to put attention to student learning behavior. Teachers also activate their knowledge about methods to gather information about students' learning and other relevant knowledge. The actual diagnosing takes place in the actional phase in a systematic manner. Predictions about students' learning processes and difficulties are made, information is gathered and interpreted to reach a concluding diagnosis. At this stage, teachers' reflections on the differences between their predictions and students' actual learning developments are useful for improving future diagnoses. In the postactional phase, a pedagogical plan derived from the diagnosis is implemented. The plan may include direct feedback to students and to parents, elaboration of individualized learning strategies or adapting the class in reaction to the diagnosis. Klug et al. (2013) emphasize that the three dimensions are related and that there is a particular connection between the postactional phase of one cycle and the preactional phase of the following cycle. The model was developed by Klug et al. (2013) first theoretically, based on a review of the relevant literature, and then validated empirically. The results of their study provided a good fit for the three-dimensional model and showed a considerable correlation between the three dimensions, which they interpreted as supporting the process characteristic of diagnosing. Moreover, they found evidence that teachers who showed a good performance on questions related to the process model were also good at achieving accurate diagnoses. Although the overall contribution of the three dimensions is small, especially relevant was the performance in the actional phase, whereas the preactional and postactional phases contributed only slightly to explain the variance in the accuracy. However, they emphasize that the aim of the model is not predicting diagnostic accuracy but focusing on the phases of the diagnostic process that lead to a more systematic diagnosis and an effective pedagogical plan that improves students' learning.

Considering different approaches and focuses on the research on diagnostic competence, Hoth (2016) points out that a comprehensive construct of diagnostic competence should consider various facets describing the diagnoses that teachers conduct in their profession. Among other facets, this would include both judgment accuracy and the diagnostic demands that teachers have to cope with in classroom situations to promote students' learning. Hoth (2016) named this latter facet "situationsbezogene Diagnosekompetenz." In a later work in English language (Hoth, Döhrmann, Kaiser, Busse, König & Blömeke, 2016), the concept of situation-based diagnostic competence was defined as "the competence that is needed to fulfill diagnostic tasks in classroom situations using informal or semiformal methods to reach implicit findings and to make adequate modifying decisions at a micro level" (p. 44). It is clear from this definition that the objects of the diagnosis are characteristics of students and instruction that may inform decisions and make the teaching and learning process more effective. Thus, students' learning and thinking processes, task characteristics and other instructional features are particularly relevant.

1.2.3   Relation of Diagnostic Competence with Teachers' Professional Competencies

Because of its relevance in teaching activities, the question of how diagnostic competence is embedded in teachers' professional competences has been investigated from different approaches.

Taking Shulman's (1987) categories of teachers' knowledge base, Leuders, Dörfler, Leuders and Philipp (2018, p. 17) suggest that diagnostic competence can be associated with the category "knowledge of learners and their characteristics" and at least one part of teachers' pedagogical content knowledge, namely their knowledge about the preconceptions and misconceptions that students may show during the learning process. Similarly, taking Weinert's (1999) approach to professional competence, Schwarz, Wissmach and Kaiser (2008) conceptualize diagnostic competence as a sub-dimension of teachers' professional competence. They also indicate that diagnostic competence requires the integration of various knowledge constituents, such as content knowledge, pedagogical content knowledge and pedagogical and psychological knowledge. However, how the different knowledge components are integrated and interact with each other remains unclear (Heinrichs, 2015; Heitzmann et al., 2019).

The COACTIV study distinguished between teachers' content knowledge, pedagogical knowledge and pedagogical content knowledge. Diagnostic competence was first conceptualized as one facet of pedagogical content knowledge, together with knowledge about representations models and explanations and about the pedagogical potential of tasks (Baumert & Kunter, 2006; Krauss et al., 2004). Later, on a revised version of COACTIV's model of teachers' professional competences, Brunner et al. (2011) state that diagnostic skills are a multidimensional competence facet requiring an integration of facets from both pedagogical content knowledge and pedagogical-psychological knowledge. The former would be particularly relevant for understanding students' mathematical thinking, knowing about common misconceptions and evaluating the pedagogical potential of tasks and their cognitive demands, whereas the relevance of the pedagogical-psychological knowledge would lie in assessment methods and the estimation of students' performance. However, Binder et al. (2018) make the claim that this allocation of diagnostic competence at both facets (PCK and PK) "remained theoretical in nature because to date no correlations of specific diagnostic skills with teachers' PCK or aspects of PK within the COACTIV data have been reported" (p. 34).

The Mathematical Knowledge for Teaching (MKT) framework (Ball et al., 2008) is another approach differentiating the knowledge needed by teachers, but it does not focus on diagnostic competence explicitly. However, other authors not belonging to this project have related its components to diagnostic activities. For instance, Philipp (2018) suggests that various domains of the MKT framework are relevant for diagnostic processes. On the one side, common content knowledge would be essential for evaluating the correctness of students' answers. Additionally, specialized content knowledge, i.e., the deeper understanding of mathematics that teachers need to make the subject comprehensible for learners, is needed in diagnostic processes for adapting task difficulties and recognizing patterns in learners' errors. On the other hand, knowledge of content and students is suggested to be crucial for diagnostic situations as it includes knowledge about common students' errors, misconceptions and preconceptions and typical students' mathematical thinking that may be essential for understanding students' reasoning and understanding of a topic.

Similarly, Ostermann (2018) regards specialized content knowledge and knowledge of content and curriculum as necessary prerequisites to evaluate the difficulty of tasks because they imply knowledge about different paths and steps to solve a task and curricular sequences that are necessary for the estimation of task complexity. He also states that knowledge about students' typical misconceptions and strategies, contained in knowledge of content and students, is a necessary component of diagnostic judgments. Moreover, in a study exploring factors that influence teachers' judgment accuracy, Ostermann (2018) highlights the crucial role of teachers' mathematical content knowledge. He explains that, although strong mathematical content knowledge is key for effective teaching, if teachers project their own knowledge into students, their judgments can be biased according to the expert-blind-spot theory. He suggests that knowledge of content and students, especially about students' misconceptions, is a significant factor for improving teachers' accuracy of judgments.

With a wider view on diagnostic activities, Leuders et al. (2018) also see a strong relation between diagnostic competence and the dimension of knowledge of content and students from the MKT framework. They state that "it comprises the anticipation of students' thinking and motivation and of their specific difficulties" (p. 17) and it "is used in situations that involve recognizing the affordances of specific content and the learning conditions and processes of particular learners" (p. 19). They also emphasize some aspects of this dimension that are closely related to the diagnostic abilities needed by teachers, such as knowledge of common students' errors for particular topics, identification and interpretation of errors, analysis of students' understandings of content and identification of levels of comprehension, knowledge about typical students' difficulties according to developmental sequences and common student strategies to solve tasks.

Following a situated approach and based on the competence model of Blömeke et al. (2015), the TEDS-FU study located teachers' diagnostic competence within the continuum model. Hoth et al. (2016) state that "teachers need specific knowledge and affect-motivational skills to diagnose students' learning during class. In these situations, they need to perceive relevant aspects and interpret them in order to decide about reasonable ways to act. Finally, the actual performance of the teacher is based on these situation-specific processes" (p. 44). In their supplementary qualitative study of TEDS-FU, Hoth et al. (2016) analyzed the situation-based diagnostic competence of teachers using their answers to the video-based instrument and their performance on the knowledge-based tests (MCK, MPCK and GPK). The focus of the tasks analyzed in this study was on teachers' perception, interpretation and decisions on classroom situations requiring diagnostic activities, in which teachers had to attend to students' learning of mathematics. Their results indicate that teachers cope with diagnostic situations very differently. In particular, they found a relation between teachers' strength in professional knowledge and the perspectives they used in situation-based diagnoses. More specifically, they state that:

> Teachers who focus on aspects of classroom management, organizational aspects and other pedagogical facets have higher pedagogical knowledge than content-specific knowledge. This may indicate that teachers with comparatively high content-related knowledge (MCK and MPCK) plan their teaching with regard to the content while teachers with comparatively high general pedagogical knowledge focus to a greater extent on pedagogical facets while planning their teaching. (Hoth et al., 2016, p. 51)

Hoth (2016) argue that diagnostic competence is a content specification of the competence as a continuum model (Blömeke et al., 2015). She explains that teachers' diagnostic competence in classroom situations can be understood as specific cognitive skills and affective-motivational dispositions that are necessary to make informal diagnostics and appropriate modification decisions during instruction aimed at fostering students' learning. Situational skills are crucial in this process because, to diagnose learning processes and students' needs, teachers have to perceive relevant situations and information, interpret them and anticipate possible strategies that may promote further learning.

The TEDS-FU study also developed a test in which teachers were asked to identify student errors under time limitation conditions (Pankow et al., 2016; Pankow, Kaiser, König and Blömeke, 2018). The identification of errors can also be seen as a component of diagnostic competence, as it is embedded within the perception component of situational skills in Blömeke et al.'s (2015) model. However, in their study, Pankow et al. (2018) found stronger relations of fast error perception with teachers' mathematical content knowledge than with mathematics pedagogical knowledge.

Based also on the model of competence as a continuum (Blömeke et al., 2015), Leuders et al. (2018) developed the concept of 'diagnostic competence as a continuum'. They argue that it comprises various approaches to understanding diagnostic competence as they focus on different parts of the model. The model specified for diagnostic competence uses the terminology of the original

and generic model of Blömeke and colleagues (2015) and is, thus, according to
Leuders et al. (2018), composed of three areas:

– Diagnostic dispositions: include cognitive facets, such as professional knowledge, and beliefs and affective-motivational facets that teachers bring into and
  activate during diagnostic situations.
– Diagnostic skills: drawing on diagnostic dispositions, situation-specific cognitive skills of perception, interpretation and decision-making are activated and
  lead to observed behavior in diagnostic situations.
– Diagnostic performance: can be understood as the product of the diagnostic
  dispositions and skills. It is the observable behavior of teachers as it occurs in
  real classroom situations. (p. 8)

1.2.4   Relevance of Diagnostic Competence for Teaching and for Teacher Education

Empirical research on teachers' diagnostic competence has notably increased in
the last years, especially in Germany but also in other countries using somewhat
different conceptualizations. The field of research on teachers' competences has
acknowledged that the level of teachers' professional competence, including general pedagogical and specific knowledge, beliefs and other motivational aspects,
has a relevant impact on the quality of instruction (see, for instance, Baumert &
Kunter, 2013; Blömeke, Kaiser, König & Jentsch, 2020; Hill & Charalambous,
2012; Hill, Blunk, Charalambous, Lewis, Phelps, Sleep, & Ball, 2008; König &
Pflanzl, 2016; Santagata & Lee, 2019).

It is widely acknowledged that diagnostic competence is particularly relevant for teaching heterogeneous classrooms. Helmke (2017) states that adapting
instruction to the needs of individual learners is of greater importance in heterogeneous groups. Individualizing and differentiating teaching strategies to meet
students' needs in learning mathematics requires teachers who can respond to
a variety of diagnostic challenges (Hoth et al., 2016). In fact, identifying learning prerequisites, recognizing different learning processes, gathering data about
individual students' abilities and understandings and adjusting instruction accordingly are considered central tasks of teaching (Bartel & Roth, 2017). Similarly,
Artelt and Gräsel (2009) state that teachers' diagnostic competence is a prerequi-
site for designing effective teaching strategies and targeted individual support and
making various pedagogical decisions, such as selecting tasks, deciding what, how

and when to give feedback to learners or developing, applying and interpreting assessments.

The emphasis on diagnosing for making pedagogical decisions is not trivial. Besides identifying the learning status of individual students, teachers have to be able to identify learning needs and choose or design strategies that promote learning (Praetorius et al., 2012). Correspondingly, Brühwiler (2014) states that the positive effect of teachers' diagnostic competence on students' learning is only possible if it is accompanied by pedagogical decisions. Diagnostic competence would provide information about the areas in which students need support, but not about the methods or strategies that would be required to promote students' learning. Moser Opitz and Nührenbörger (2015) also point out that the connection between diagnosis and the provision of adapted learning opportunities is complex. Diagnosis should enlighten pedagogical interventions by providing information on the specific areas in which a student needs support. However, they claim, the extent to which targeted support can be developed based only on the diagnosis remains unclear. Both support and diagnostic strategies need to be based on pedagogical and subject-specific considerations. Similarly, Götze, Selter and Zannetin (2019) state that intervention strategies without a previous diagnosis are too unspecific and a diagnosis disconnected from supporting strategies is also ineffective. In other words, teachers should develop and use continuous diagnostic strategies or instruments not only for collecting information but also with the aim of informing the process of designing and implementing pedagogical interventions. Therefore, Götze et al. (2019) recommend the use of targeted and informal diagnostic procedures that can be continuously integrated into everyday teaching. These may include, among others, informal observation of students during lessons and of their written work.

Continuous diagnosis is also emphasized by Ashlock (2010), who suggests that mathematics teachers should be continually assessing. He remarks that continuous assessment is different from continuous testing, as information about student learning can be collected from different sources. The information gathered should be used to follow up on students' learning progress, to improve teaching strategies and guide instructional decisions. As learning is an individual process, data collected about individual students' knowledge and thinking can be used to adapt lesson plans and make ongoing decisions for teaching particular students. Therefore, diagnosis and instruction are combined in a cycle in which observation and understanding of student learning are followed by adapted instruction, leading to a repetition of the cycle. Each cycle can occur very quickly and many times in one lesson or occur throughout several lessons. In other words, diagnostic teaching requires careful observation aimed at identifying the mathematical ideas and procedures that students are, in fact, using. If they are not correct, targeted intervention is called for. The interpretation of the collected data leads to thoughtful inferences about student learning, constituting the basis of teaching decisions. Diagnosis is, thus, a means for providing differentiated and individualized learning opportunities for students.

Due to the observed importance of diagnostic activities for teaching, a number of empirical studies have explored the relationship between the diagnostic competence of teachers and student learning. One approach has been investigating the influence of the diagnostic competence of teachers, conceptualized as accuracy of judgment, on students' performance gains. In a study examining the relationship between the three accuracy components and student achievement, Schrader and Helmke (1987) found that diagnostic competence alone produced none significant effect on students' learning gains. Only when structuring support was given during a lesson, teachers' diagnostic competence played a significant role. Structuring support, provided in the form of prompts and remarks emphasizing or pointing to relevant information for solving tasks, needs to be adapted to the learning status and needs of students. The greater learning gains were found when a teacher's high diagnostic competence was complemented with a frequent use of structuring supports in the classroom. The students of teachers with a weak diagnostic competence showed smaller learning gains, regardless of the frequency of the structuring support. Probably because structuring support is most helpful when adapted to the learning needs of students. The least learning gains were shown by students with teachers displaying a strong diagnostic competence and who rarely gave structuring support. This unexpected result is interpreted as an indication that a high diagnostic competence can even be unproductive for students' learning if teachers do not have the ability to connect their judgment of student learning with pedagogical strategies. Students can then become aware of their difficulties but are not provided with the opportunities to work on them. Thus, their results suggest that diagnostic competence is not a sufficient condition for effective teaching, but it is a necessary requirement for optimizing particular teaching strategies so that they successfully foster learning.

Within the COACTIV study, the relation between mathematics teachers' diagnostic skills, the quality of their instruction and students' performance was examined (Anders, Kunter, Brunner, Krauss & Baumert, 2010). They found a significant relationship between teachers' ability to accurately judge the difficulty of tasks and the quality of their instruction, operationalized as the cognitive activation potential of the mathematical tasks they included in class tests. On the contrary, they found no significant correlation between diagnostic sensitivity or rank component, i.e., teachers' accuracy in appraising a rank order of students' performance, and the cognitive quality of their test tasks. Furthermore, they found that both indicators (teachers' accuracy of judgment of task difficulty and teachers' accuracy in estimating a rank order of students' performance) were related to students' achievement. Albeit small, these relations were interpreted as relevant because of the complexity of the teaching process and the multiple factors that also play a role in students' achievement gains. In other words, when controlling for background and context conditions, students demonstrate better performance in mathematics when their teachers are more skillful in accurately judging the difficulty level of tasks and the performance rank of students in their class (Brunner et al., 2011). Thus, the results were considered relevant for supporting the importance of teachers' diagnostic competence in the classroom (Anders et al., 2010). Surprisingly, in a later analysis, negative effects of teachers' judgment accuracy of student achievement on the cognitive quality of the tasks they included in class tests were found (Binder et al., 2018). The authors explained that the lack of accuracy on teachers' judgments of their students' achievement may provoke teachers venturing to include more cognitively activating tasks in their class tests. Additionally, they found that teachers' accurate judgment of the correct solution rate of their students on specific tasks has a positive impact on students' mathematics achievement, arguing again in favor of the relevance of teachers' diagnostic competence on student learning.

From a different approach, the LMT project investigated the relationship between teachers' mathematical knowledge for teaching and student achievement in mathematics (Hill et al., 2005). Their conceptualization of mathematical knowledge for teaching includes, besides diagnostic competence, all the skills and knowledge required for the work of teaching. The gains in mathematics achievement of 1190 first-grade and 1773 third-grade students over one year were related in a nested model to the knowledge of 334 first-grade and 365 third-grade teachers. Their results indicate that, after controlling for relevant teacher and student background covariates, student's gains in mathematics achievement can be predicted by their teachers' mathematical knowledge for teaching both for the first- and third-grade cohorts. Results also show that teachers' mathematical knowledge for teaching was a stronger predictor than any other teacher background variable and daily time spent on mathematics instruction. In a more recent study, using data from 284 teachers, Hill and Chin (2018) found evidence showing that teachers' judgment accuracy is an identifiable and stable teacher trait and that teachers' performance on this metric was related to their mathematical knowledge and predicted student outcomes. Moreover, their analyses provided evidence of a main effect of teachers' knowledge of students on students' learning outcomes.

1.2.5   Fostering Diagnostic Competence in Initial Teacher Education

Because of the increasing acknowledgment of heterogeneity in school classrooms and the importance of diagnosing and fostering students' learning by adapting teaching to individual needs, the development of preservice teachers' diagnostic competence already in initial teacher education has been recognized as highly relevant (Artelt & Gräsel, 2009; Brandt, Ocken & Selter, 2017). Götze et al. (2019) suggest that the use of informal diagnostic methods integrated during everyday teaching, such as the observation of students while they work and collection of information from their written tasks, is most productive when teachers are trained or have experience on quickly looking at details, asking specific questions and understanding students' explanations. Moreover, because knowledge influences what teachers perceive and how they interpret what they observe, knowledge about learners' mathematical thinking that may act as a base for diagnosis is also considered helpful.

In a study examining preservice teachers' analyses of students' conceptual understanding of mathematics, Bartell, Webel, Bowen and Dyson (2013) found that content knowledge is a necessary but not sufficient condition for supporting their evaluation of students' thinking. Thus, they advocate for providing opportunities for preservice teachers to engage in such evaluations from early on within teacher education programs, simultaneously to the development of their professional content knowledge. Similarly, Bartel and Roth (2017) suggest that the development of preservice teachers' diagnostic competence should ideally occur simultaneously with the development of their pedagogical content knowledge. By applying theoretical knowledge in classroom-related situations, especially to make diagnoses of students' learning processes, preservice teachers can make connections between theory and practice, develop their diagnostic competence and understand the relevance of theories for their profession. Brandt et al. (2017) also suggest that the development of the competence should occur continuously during teacher education and in close relation to subject-specific courses. The use of case studies and simulation experiences could be helpful learning opportunities. With the support of authentic materials, such as transcripts or videos of classroom situations and samples of students' written work, real classroom situations and students' comprehension processes can be analyzed and discussed without the pressure of making the on-the-moment decisions that teaching requires.

Besides the knowledge component, Barth and Henninger (2012) emphasize the role of situational cues. They developed a multimedia learning environment aimed at developing preservice teachers' skills in making competent diagnoses in teaching situations. It included videos of classroom situations and encouraged preservice teachers' conscious reflection on interpretations of students' understandings. Their results indicate that providing realistic and contextualized situations, accompanied by detailed information about the teaching lesson, was crucial because preservice teachers mainly used this information to make their diagnoses, in contrast to the use of previous knowledge about the class. Moreover, situation-specific information supported preservice teachers in reflecting on their own interpretations of student thinking and becoming aware of their diagnostic behavior.

Videos have been commonly used in teacher education programs for developing preservice teachers' professional competences (see, for example, McDuffie et al., 2014; Santagata & Guarino, 2011; Star & Strickland, 2008; van Es, Cashen, Barnhart & Auger, 2017). As an artifact in university courses, videos allow for whole-class discussions of teaching strategies, student thinking and other issues related to teaching and learning, as everyone is observing the same situation. Additionally, videos can be paused or shortened to direct preservice teachers' attention to specific aspects or moments. In fact, Sherin and van Es (2005) found that watching and discussing classroom videos can support preservice teachers in learning to focus on significant classroom events rather than identifying all of the events in a video as relevant. They also found that videos can be supportive in shifting preservice teachers' analyses of videos from descriptions followed by judgments of the teaching situations to a more interpretative stance, in which they intended to understand the influence of pedagogical decisions on students' learning. Moreover, preservice teachers' comments about what they observed were increasingly supported by evidence from features or events from the videos.

In a similar line of research, Star and Strickland (2008) included a systematic use of videos in a teaching methods course with the explicit goal of improving observation skills. Besides giving extensive opportunities to watch and discuss classroom videos, the course provided preservice teachers with an observation framework that served to guide their attention. Results indicate that the course supported the development of preservice teachers' observation skills and led them to notice relevant features of classroom situations. Santagata et al. (2007) also implemented a video-based course for preservice mathematics teachers. Unlike other studies, their course used videos of entire lessons, without editing. As Star and Strickland did, they provided preservice teachers with an observation framework, directing them to make connections between learning goals, teaching strategies and students' learning. After the course, preservice teachers were able to give more specific and elaborated comments that focused more on mathematical content and student learning. Similarly, Santagata and Guarino (2011) found that video-based activities supported the development of preservice teachers' ability to notice specific aspects of student thinking and the strategies used by teachers to guide student learning.

However, the incorporation of video-based activities itself does not assure the development of teacher competences. Jacobs and Philipp (2004) point out that the usefulness of analyzing students' work in teacher education is not determined by the work samples or the events selected but on the discussions that can be generated from them. They are a useful artifact as long as they facilitate discussions about specific details of students' thinking, related mathematical content and teaching strategies. For supporting preservice teachers noticing of students' mathematical thinking, Jacobs and colleagues (2010) provide prompts that may aid teacher educators in generating productive discussions. They argue that preservice teachers have difficulties in attending to children's strategies, so they need directed support to shift their focus on general issues towards specific features of students' understandings, mathematically relevant aspects of the strategies they use and significant details in the explanations they give. They also suggest that to interpret students' understandings, besides attention to clues of student thinking, a certain level of mathematical knowledge is needed. Thus, preservice teachers who avoid interpreting students' mathematical understandings may be lacking sufficient mathematical knowledge and calling for support in making sense of students' strategies and connecting them to mathematical concepts and procedures. Finally, they advise that, although preservice teachers' instructional decisions or proposals may vary widely, they are more likely to be productive when they consider student thinking. So, a shift towards detailed and concrete pedagogical suggestions that include specific tasks considering students' existing understandings and anticipation of future strategies should be encouraged.

Van Es et al. (2017) also point out that preservice teachers need structured guidance to benefit from the use of video segments. They investigated the development of preservice teachers' abilities to notice ambitious mathematics practices in a video-based course. In their analysis of the results, they considered that "careful selection of the videos, along with structured guidance in viewing the clips, appeared to help orient the candidates' attention to the mathematics and student thinking, and to the relation between teaching moves and student thinking and learning" (van Es et al., 2017, p. 182).

Also from a situated approach but for a different purpose, Blömeke et al. (2015) draw attention to the usefulness of videos in the context of competence assessment. They suggest that videos may complement other traditional forms of measurement, such as multiple-choice and constructed-response items, by allowing the inclusion of real-life situations. Besides cognitive and affective dispositions, competence assessment should incorporate the measurement of situational skills and observable behavior. Videos were seen as a useful tool for bringing representative real classroom situations, that consider the complexity and lack of structure of schools, into competence assessment. In line with this suggestion, videos have been used in several studies of teachers' professional competences (e.g., Heinrichs, 2015; Hoth et al. 2016; König, Blömeke, Klein, Suhl, Busse & Kaiser, 2014).

Kaiser and colleagues (2015) discussed the theoretical and methodological challenges of video-based assessments for the evaluation of teachers' competence from a broad perspective. They report that the use of video does enrich the traditional testing items by making it possible to identify how teachers perceive, interpret and decide upon given classroom situations. Evaluating teachers in a holistic way also permits distinguishing between expert and novice teachers because the video-based items show particular teaching events for which expertise is needed, for instance, by requiring the integration of different types of teaching knowledge. However, they also recognize some limitations. Although videos are closer to real classroom situations, they are still not completely real and incorporate some shortcomings into measurement. For example, the selection of relevant events or features is, to a certain extent, made by the camera and the video edition, making it difficult to evaluate perception skills. Another limitation is related to the decision-making process. When asking teachers for their pedagogical decisions after a video, they have to write them down, which occurs with a different timing and reflection process than it would occur in a real classroom and the need for on-the-moment decisions. Despite these limitations, the authors conclude that videos can notably enhance the assessment of teachers' professional competence.

1.3   Diagnostic Competence in Error Situations

1.3.1   Role of Errors in Mathematics Teaching and Learning

Errors are inherent to the process of learning and, thus, are expected to be encountered in mathematics learning as well (Fiori & Zuccheri, 2005; González López, Gómez & Restrepo, 2015; McGuire, 2013; Oser, Hascher, Spychiger, 1999; Radatz, 1980). Educational theories envisage errors in markedly different ways (Santagata, 2005). On the one side, the behaviorist approach envisions errors as something to be avoided. Students learn by successive positive contact with the content or behavior being learned. Thus, exposure to errors should be evaded to limit the chances that a wrong answer is reinforced. On the other side, constructivist learning theories suggest that learners build new knowledge by using existing cognitive structures to actively interpret and understand environmental input (Smith, diSessa & Roschelle, 1993). Hence, errors and misconceptions play an important role in constructivist approaches to learning and are generally interpreted as signals of an inconsistency between previous cognitive structures and new knowledge (Brodie, 2014; Santagata, 2005; Smith et al., 1993). In other words, errors found in students' work show flaws in their understanding of concepts in the way adults or science do (Radatz, 1980). Brodie (2014) states that errors "arise from misconceptions, which are conceptual structures constructed by learners that make sense in relation to their current knowledge, but which are not aligned with conventional mathematical knowledge" (p. 223).

Errors play a crucial role in the identification of students' faulty conceptions in the classroom and inform the adaptation of teaching strategies (Santagata, 2005). Already in 1975, Cox emphasized that "it is important to identify the student making systematic errors because this type of error is potentially remediable" (Cox, 1975, p. 152). For procedural skills, Brown and Van Lehn (1980) explain that the focus needs to be on the detection of the procedure that is generating the erroneous answers. Consequently, teachers' ability to recognize a student's misconception by analyzing the available evidence on their erroneous answers is central to the work of teaching (Brown & Burton, 1978). Rather than re-teaching concepts, teachers should develop strategies to dealing with students' errors based on their understandings of the interconnections of ideas and concepts that students are making and the identification of the connections in which errors and misconceptions arise (Brodie, 2014). This implies that teachers need to have the ability to identify and interpret students' errors and misconceptions, which is part of teachers' diagnostic competence.

1.3.1.1 Conceptualization of Errors

Before getting deeper into the role of the diagnosis of errors for making instructional decisions, the concept of errors needs to be clarified. Cox (1975) differentiated between three types of errors in the computational domain, namely systematic errors, random errors and careless errors. Random errors occur often but present no recognizable pattern of wrong answers. Careless errors are encountered less frequently and can be attributed to a simple distraction without evidence of lack of understanding. Systematic errors can also be found in students' work and they display a pattern of wrong answers. Identifying the pattern in the errors made by a student is necessary for selecting a remediation strategy. Similarly, Radatz (1980) described student errors as systematic when referring to error analysis for mathematics learning. He regards errors to be systematic and the consequence of previous learning experiences. As errors reveal student's flaws in the understanding of concepts and procedures, they are also persistent, i.e., they will continue to occur unless pedagogical actions are taken to remediate them. Therefore, he emphasized the importance of error analysis as an opportunity to locate the difficulties students are having in the learning process and design appropriate and individualized support to promote student understanding.

Although a wide variety of terms has been used to refer to students' erroneous conceptions, the most commonly employed has been the notion of misconceptions (see Smith et al., 1993). Misconceptions are described as a sequence of reasoning based on an incorrect underlying idea (Brodie, 2014; Nesher, 1987; Smith et al., 1993). Thus, misconceptions produce systematic errors because learners apply the same incorrect guiding rule. In contrast, errors that occur randomly and cannot be attributed to the application of wrong principles are considered as mistakes and called like this (Holmes, Miedema, Nieuwkoop & Haugen, 2013; Nesher, 1987). The cognitive structures associated with misconceptions are built by students founded on earlier learned structures as students overgeneralize or simplify prior knowledge and use it in new domains, in which they are actually not correct (Nesher, 1987; Smith et al., 1993). Most often, misconceptions are extensively spread among learners and can be consistently found across contexts, suggesting that they are not the result of particular learner or teacher characteristics (Brodie, 2014; Smith et al., 1993). Moreover, misconceptions are regarded to be persistent and difficult to modify because they make sense for the student, according to their knowledge (Smith et al., 1993). For this reason, the role of teachers in creating the learning conditions so learners can restructure their understandings is very complex (Brodie, 2014). Despite this difficulty, misconceptions have to be identified, approached and taken care of during teaching because they interfere with the learning process (Smith et al., 1993).

Ashlock (2010) also differentiates between careless mistakes and misconceptions. He focuses on the origin of misconceptions and states that the way students learn correct and erroneous mathematical ideas and procedures is the same. Students attend to some features that new information appears to have in common and make connections with their own previous knowledge. Often, initial experiences that do not include enough examples or variety may lead to building an erroneous concept or making wrong connections. In addition, individual students do not possess all the required previous understandings and skills to learn new mathematical ideas and procedures and thus, are not able to make the intended correct connections and they construct their own (incorrect) rules to make sense of new concepts. When students apply these incorrect understandings, they produce error patterns. These systematic errors are evidence of the misconceptions students have constructed. Hence, Ashlock (2010) argues that, besides evaluating if results of computation procedures are correct or incorrect, teachers have to be alert for error patterns while examining students' work so they can recognize how individual students are thinking and modify instruction accordingly.

In the German context, the early work of Weimer (1925) distinguished between *Fehler* (error) and *Irrtum* (mistake). According to Weimer, errors are unintentional inaccuracies that result from a failure of psychological functions such as lack of attention or concentration. Mistakes, in turn, are the result of a state of lack of correct information, which cannot be attributed to a failure of psychological functions (Weimer, 1925; Heinze, 2004). In addition, central in German research about dealing with errors in teaching situations has been the errors' theory from Oser and colleagues. They define errors as facts or processes that deviate from the norm, which represents the reference system. Without norms, it would not be possible to distinguish between right and wrong, i.e., to identify what constitutes an error (Oser et al., 1999). This definition also suggests a critical positive role for errors in the learning process, namely that they are useful in learning to differentiate correct from incorrect facts and procedures. Heinze (2004) specifies this definition for mathematics teaching and learning and states that errors are claims that differ from generally accepted statements, methods and definitions of mathematics. In this sense, errors play an important role in the learning process, they help to build negative knowledge, to sharpen the understandings of what is correct and what is incorrect, what is true and what is false (Heinze, 2005).

In the present study, the definition from Heinze (2004) is used, considering that errors in the mathematics classroom are statements or procedures that are different from generally accepted mathematical ideas and processes. Particularly interesting for this study are systematic errors, namely those errors that show a particular pattern and provide evidence of an underlying misconception, an incorrect understanding of mathematical concepts and/or procedures (Ashlock, 2010).

Besides the differentiation between systematic and non-systematic errors, a variety of other classifications of errors has been developed. Some approaches have focused on finding, characterizing and grouping types of errors in mathematical topics, while others differentiate types of errors according to their causes. In what follows, error classification approaches relevant to the interests of the present study will be described and discussed.

Malle (1993) describes causes of errors in the areas of arithmetic and algebra very clearly and differentiates between errors in information acquisition and in information processing. Errors in information acquisition include, for instance, the use of an incorrect arithmetic operation by choosing one that is easier to calculate and has been previously learned. Contributing to this error is the selective and incomplete perception of elements and the connections that students build between certain school activities and the mathematical operations they have to use. For example, when in primary school, students have to solve several addition tasks, they see two numbers with another sign between them and they automatically sum them up. When later on, other operations are learned, students also perceive the operation symbol incompletely and relate it to either an increase or a decrease, also adding in the multiplication tasks and subtracting in the division tasks. According to Malle (1993), these difficulties often arise when many tasks of the same type and rarely a mixture of task types are included in instruction.

The second error type described by Malle (1993) refers to errors in information processing. This includes, for example, when students develop erroneous conceptualizations by overgeneralizing procedure rules outside their valid scope. In addition to overgeneralization, the use of inadequate schemes can also lead to errors. Inadequate schemes are those, which are generally incorrect but work for some particular cases; they are not taught but are usually developed by students. Furthermore, Malle (1993) suggests that students sometimes spontaneously develop demand schemes, which they do not store permanently but form and apply in a situation-specific manner. Recurring to general daily-life knowledge that is unrelated to mathematics or using too open and unspecific schemes or inappropriate substitution schemes can also lead to students' mathematical errors. Even more, Malle (1993) suggests that certain tasks can elicit the activation of more than one scheme, and if the student does not choose the correct one, the schemes can interfere with each other.

Using general life skills that are not related to mathematics or using too open and unspecific schemes or inappropriate substitution schemes can also lead to student mathematical errors. Furthermore, Malle (1993) suggests that certain tasks can cause more than one scheme to be activated, and if the student does not choose the correct one, the schemes can interfere with each other. Another common cause for errors suggested by Malle (1993) is a lack of attention to the hierarchy of procedures, especially when complex operations are involved at different levels. Finally, Malle (1993) points out that even when students choose the correct scheme, its application can lead to execution errors, which are commonly considered processing errors or careless mistakes and can be quickly recognized and corrected by the students.

More recently, Hansen, Drews, Dudgeon, Lawton and Surtees (2017) suggest that misconceptions, i.e., those understandings that differ from accepted mathematical ideas and lead to errors in students' work, can be the result of the incorrect application of a rule, the overgeneralization or the undergeneralization of certain mathematical conceptions. Ashlock (2010) describes overgeneralization misconceptions as those occurring when conclusions are drawn before enough information is available for the learner and concepts, rules or procedures are applied in contexts in which they are not adequate. Overspecializing misconceptions are generated when rules are applied incorrectly, restricting the procedures or contexts of application inappropriately.

A different classification of causes of errors was derived by Holmes et al. (2013) after analyzing over 300 teachers' evaluations of students' errors. They discriminate between vocabulary misconceptions, computational errors and erroneous beliefs or misconceptions. Vocabulary misconceptions refer to an incorrect use of mathematical terminology, they focus on language errors. Computational errors are those found in calculation facts or procedures. Although they can be difficult to distinguish from erroneous beliefs, they show no systematic patterns. Erroneous beliefs, which they also associate with misconceptions, are inaccuracies in mathematical thinking, they imply a conceptual error.

Computational errors and conceptual inaccuracies are also included in the classification of Götze et al. (2019). They distinguish two types of error patterns focusing on written computational algorithms. On the one side, errors in recalling basic addition and multiplication facts and facts with zero and, on the other side, understanding errors. The latter are classified into three further categories. One refers to errors related to regrouping and renaming in addition and subtraction procedures. A second category refers to errors resulting from difficulties in the understanding of the operation. The third category includes errors resulting from a lack of understanding of place value issues.

A further categorization of errors has been developed by Scherer and Moser Opitz (2012). They distinguished and exemplified the five types of errors suggested by Jost, Erni and Schmassmann (1992). These include *Schnittstellenfehler,* which are errors by the perception, reproduction and notation of mathematical symbols that arise due to sensory problems, such as auditive, visual or spatial orientation difficulties. Additionally, they also describe understanding errors of two different types: errors in the understanding of concepts and errors in the understanding of operations. The former refer to errors due to a failure in recognizing relationships and comprehending concepts, such as place value or the concept of fractions. The latter are described as an insufficient comprehension of what operations are about or a limited comprehension of their function, for instance, when students do not understand that in subtraction something is taken away from a total and that it also can be used when to determine how much is needed to complete a greater quantity, or that in division a number is distributed in equal parts until no further equal quantities can be generated. The fourth type of error is automatizing errors, which occur despite the correct understanding of concepts and operations, affecting processes that cannot be automatized. For example, when instead of recalling a basic addition fact, students use the counting-on strategy and they also count the number they start with, the result is one number smaller than it should be. Errors that are a consequence of perseveration fall into this category as well. They arise, for instance, when the first four tasks are addition, the fifth question is a subtraction and the student continues adding. The last category describes application errors that arise when mathematical concepts and operations are transferred to a new context, in which they cannot be used correctly. Scherer and Moser Opitz (2012) make it clear that for each error observed in a students' work, it is possible to generate multiple hypotheses, which relate it to different categories of error types. In other words, the same error could have an underlying cause that can be attributed to more than one of the mentioned categories. Hence, they emphasize that it is important to collect as much information as possible during the analysis of student work by looking not only at the incorrect answers but at the correct ones too and by evaluating if it is or not a systematic error. Distinguishing between situations in which an incorrect procedure is used and those in which the student does not know what to do is also valuable information to understand their thinking (Ashlock, 2010). Despite the classification of the errors, broadening the spectrum of possible causes and collecting evidence of their thinking is crucial for teachers to understand their students' mathematical reasoning.

1.3.1.2 Role of Error Diagnosis in the Classroom

Error analysis can be a useful diagnostic instrument that can be practically integrated into instruction (Scherer & Moser Opitz, 2012). By analyzing students' errors, teachers can improve their understanding of individual students' thinking and adapt their teaching strategies accordingly. The identification of an error during a lesson should provoke the teacher to make some adaptations to the situation. It should trigger an instructional reaction that provides the student with feedback or hints aimed at promoting further learning. Thus, it is possible to state that a key goal of error analysis is the modification of the teaching and learning process in order to account for individual learners' needs.

Ashlock (2010) suggests that teachers should teach diagnostically, meaning that careful observation of students' work is needed, so misconceptions and error patterns can be identified. Recognizing the different computation strategies that students develop is important to understand how each student is thinking and, when error patterns arise, teachers can hypothesize about possible causes and think about alternatives for targeted teaching. Scherer and Moser Opitz (2012) also emphasize the importance of the process of developing hypotheses about the type of error and its underlying causes. They argue that it is crucial to use any instruments or strategies that prove to be useful for confirming those hypotheses, so relevant information can be gathered and taken as the base for making pedagogical decisions. This makes evident that the end goal of identifying the causes of students' errors is not just understanding students' thinking but to respond in a way that promotes learning.

In order to confirm (or reject) their hypotheses about students' understandings or misconceptions, teachers need to make use of various methods to collect supporting evidence. Because most of the errors teachers observe arise during complex classroom situations, in which teachers have to consider multiple aspects simultaneously, pay attention to a diversity of students and make decisions rapidly, semiformal and informal methods for collecting information are preferred (Heinrichs, 2015). During instruction, there is seldom time to apply structured or standardized tests to diagnose misconceptions. Instead, other strategies that are always available, such as small interviews or observations, can be used spontaneously and on-demand. Individual interviews are a good method for collecting information about the thinking of a student (Ashlock, 2010; Holmes et al., 2013; Radatz, 1979; Reinhold, 2015). Small talks between the teacher and a student during individual work may provide valuable insights into students' understandings of mathematical ideas and procedures if students are encouraged to answer with as much detail as possible and to think aloud while they solve a task and asked to explain the reasons for their answers and their choices. Radatz (1980) alerted for the role that the process of questioning and the interaction between teacher and student may play on student thinking. Students obtain clues from the questions themselves and from teachers' gestures and usage of language, which may influence their responses. Careful and attentive observation of students during interviews and during teaching and learning situations is crucial to understand their thinking and identify errors and misconceptions. According to Ashlock (2010), relying on observation procedures to capture students' thinking is not as easy as it may seem, because when students have difficulties understanding a topic, they usually develop strategies to hide their confusion, so it appears that they correctly understand the mathematical ideas even when they do not. For this reason, he advises the use of various observation methods, ranging from informal, through interactive, to structured, and combining them with interviews and careful listening of student ideas. Similarly, Radatz (1980) pointed out that the observation of students while they are solving mathematical tasks, as well as the analysis of students' written work, diagnostic interviews and think-aloud interactions, are useful and easily applicable methods to analyze students' errors. They are methods that are always available and are flexible to be used in various situations where hypotheses about the causes of students' incorrect understandings need to be made.

It is clear that analyzing students' work beyond scoring correct and incorrect answers can be a rich source of information about student thinking and learning. The identification of error patterns can provide teachers with valuable insights about the aspects and ways in which students' mathematical thinking requires further development and, hence, inform teaching decisions (Hansen et al., 2017). Moreover, there is evidence showing that dealing with errors in the learning process is more effective for the development of new knowledge and skills than trying to avoid them (Keith & Frese, 2008). However, for error situations to be used productively, there must be a positive error culture in the classroom (Spychiger, Oser, Hascher & Mahler, 1999). Discussion of errors are often interpreted by learners as unpleasant situations and the negative emotions associated with making errors can obstruct the potential that they may have in the learning process (Rach et al., 2013). Additionally, teachers also tend to give a negative connotation and to avoid student errors in the classroom (Oser et al., 1999; Scherer & Moser Opitz, 2012). On the contrary, in error-tolerant classroom cultures (Rach et al., 2013), errors are allowed, acknowledged and accepted and when they arise, they are not avoided or negatively evaluated but used as learning opportunities (Spychiger et al., 1999). Ashlock (2010) points out that in order for students to cooperate in the diagnosis of their difficulties, they have to perceive that their responses are well accepted, even when they are not correct. If students are afraid of making mistakes, they do not take any risks during the learning process, which can hinder the acquisition of further knowledge and the development of further skills. Thus, a classroom atmosphere showing a positive attitude towards errors is important, so students dare to make errors during the learning process, knowing they will be taken seriously and supported both by their teachers and other pupils (Götze et al., 2019).

1.3.2   Teachers' Diagnostic Competence in Error Situations

According to Prediger and Wittman (2009), situational identification and handling of errors during instruction pose a great challenge for teachers and their diagnostic competence. In the midst of complex classroom situations, they have to evaluate if an error is a simple slip or if it is the result of a misconception, and then they need to ponder the relevance of the error for the goal of the lesson to decide if it is worth deviating the planned activities. If they decide addressing the error is relevant for achieving the objective of the lesson, they have to consider either interrupting the ongoing discussion or paying attention to it later. Additionally, they have to think about alternative support strategies and choose the most appropriate option to modify their instruction.

Heinrichs (2015) developed a definition of diagnostic competence in error situations that she used as the basis for her study with secondary preservice teachers. Her definition will be used in the present study as well. Thus, the diagnostic competence teachers require when errors arise in the classroom will be understood as:

> the competence that is necessary to come to implicit judgements based on formative assessment in teaching situations by using informal or semiformal methods. The goal of this process is to adapt behavior in the teaching situation by reacting to the student's error in order to help the student to overcome his/her misconception. (Heinrichs & Kaiser, 2018, p. 81)

It is worth noticing that the definition uses the construct of competence. Hence, it implies more than only knowledge or skills. It considers both cognitive and affective-motivational dispositions, as well as situation-specific skills, as proposed by Blömeke et al. (2015; see section 1.1.3).

1.3.2.1 Research State of Teachers' Diagnostic Competence in Error Situations

A decade ago, Son and Sinclair stated that "preservice teachers' responses and their strategies to student errors is an area that has received limited attention in the research literature" (Son & Sinclair, 2010, p. 31). However, studies exploring teachers' and preservice teachers' interpretation and handle of errors in mathematics teaching situations have increased in recent years. Results from studies focusing on teachers' competence to identify, interpret and analyze student errors as well as using this information to make instructional decisions will be presented in the following.

In a qualitative study analyzing 86 assignments collected from preservice teachers in three different middle-school and elementary methods courses, Cooper (2009) explored how preservice teachers analyzed samples of students' work. After analyzing some student errors during the university course, preservice teachers received a homework assignment in which they were asked to identify the computational error pattern in the work samples, describe the possible misconceptions underlying the error patterns and recommend possible instructional strategies to support the students overcoming their errors. Results indicated that although all the preservice teachers were able to identify computational error patterns correctly in each sample of student work, 80% provided a reasonably possible rationale for the misconception explaining the errors. Greater difficulties were identified in the proposed instructional strategies for remediation. A majority (67%) of the responses focused on re-teaching with a focus on procedural methods, such as reviewing rules or a step by step explanation in which the teacher would show the procedure.

Seifried and Wuttke (2010a) also identified teachers' difficulties in dealing with student errors that occur during lessons. In a pilot study exploring video-recorded lessons from two experienced teachers, 76 error situations were selected. Results indicated that teachers do not always handle student errors constructively and do not adapt their strategies systematically to the characteristics of the error situation. Moreover, reasons for the errors were rarely recognized and the feedback provided to students showed a low elaboration level. These results were considered only as preliminary due to the rather small sample size. Therefore, in a subsequent study, the research group examined the development of teachers' competence to diagnose errors and deal with them by testing them at different stages of their teaching development (Türling, Seifried & Wuttke, 2012; Wuttke & Seifried, 2013). First, they identified typical errors for the domain of accountancy by interviewing experts about error situations and typical specific errors (Seifried & Wuttke, 2010b). They used these data to produce video vignettes presenting short classroom error situations, which were used as prompts to test teachers' professional error competence. In the context of an interview, teachers and prospective teachers were asked to identify errors and describe how they would respond to them (Türling et al., 2012). In addition, they used a paper-pencil-test, in which participants had to identify and correct errors in a students' assignment, and a self-perception questionnaire on professional error competence (Seifried, Wuttke, Türling, Krille & Paul, 2015). Data from 287 German teachers and prospective teachers in different stages of their training were collected: university students in bachelor and master stages, preservice teachers in their practical training stage and in-service teachers. Based on performance in both the video vignette and the paper-pencil tests, their findings reveal that, in contrast to in-service teachers, students and preservice teachers showed a low ability to identify and correct student errors (Türling et al., 2012; Wuttke & Seifried, 2013). Regarding strategies for handling errors, results indicate that in-service teachers outperform bachelor- and master-students and preservice teachers in quality measures, such as cognitive activation of learners or providing an elaborated reasoning for the strategies they decide to use in error situations (Seifried et al., 2015). Moreover, they found that student-teachers and preservice teachers tend to overestimate their own ability to diagnose errors, whereas professional teachers apparently see their ability in a more realistic way (Türling et al., 2012). In summary, their studies found deficits in prospective teachers' abilities to identify errors and handle them appropriately to foster student learning.

Differing from the positive findings of Seifried et al. (2015) regarding in-service teachers' abilities to handle student errors, Riccomini (2005) found that although teachers were able to identify specific error patterns, they did not focus their instructional strategies on them. He investigated the ability of 90 general and special education elementary teachers to identify two types of error patterns in subtraction in samples of student work. Analyses of participants' responses indicate that 60% of the teachers were able to correctly identify and describe both specific error patterns in the students' work. Additionally, the data suggested that teachers' instructional focus was seldom on the pattern of errors. Even a great majority of the teachers who could correctly identify the error pattern did not base their instruction on it. Most often, teachers decided to target their strategies towards re-teaching basic subtraction facts. Only 31% of the participant teachers chose an appropriate strategy for both error types. This makes it evident that the relationship between error identification and decision-making is not straightforward and that designing appropriate instructional strategies poses the greatest challenges.

In a case study, Son and Sinclair (2010) identified that even though preservice teachers recognized a student error in terms of conceptual aspects, they tended to rely on procedural knowledge and to use strategies based on "showing and telling" when responding to the student's error. The researchers attributed this difficulty to an insufficiently developed pedagogical content knowledge. In a later study, Son (2013) examined the interpretations and responses of 57 preservice teachers to a student error. Although the error exhibited a strong conceptual origin, most preservice teachers attributed it to procedural aspects and suggested procedural guidance for the student. Preservice teachers who showed difficulties with the mathematical knowledge themselves were particularly likely to provide inappropriate interventions. However, preservice teachers with sufficient mathematical knowledge did not necessarily give suitable responses. In many cases, even though preservice teachers understood the mathematical topic conceptually, they interpreted the error as a procedural one and suggested a strategy based on telling the student the right procedure, without providing the opportunity for the student to reflect on her method. Son (2013) suggests that the disconnection between preservice teachers' understanding of an error, its interpretation and their pedagogical responses may be explained by their insufficiently developed mathematical and professional knowledge, by their own experiences learning mathematics in a procedural way and by their resistance and difficulty to teaching mathematics for understanding.

Examining the influence of knowledge on the handling of errors, Bray (2011) conducted a case study with four in-service teachers. She focused on how these teachers responded to student errors during classroom discussions and how knowledge and beliefs influenced their error-handling strategies. Findings of the study suggest that teachers' beliefs are closely related to the decisions of including or not the analysis of incorrect solutions and procedures during classroom discussions. For one of the teachers, including the analysis of errors in whole-class discussions was essential to build solid mathematical knowledge. However, the other three teachers expressed the concern, also found by other studies, that this kind of discussions may obscure the understanding of some students, especially from those with a more fragile mathematical basis, and that exposing the error of one student to the whole class may embarrass that child. This approach relates to a teacher-centered behavioral paradigm in which mistakes must be avoided and, when they occur, the correct procedure and outcome must be pointed out to students (Nolte, 2021). Regarding the role of knowledge, results suggest that it seems to determine how discussions about errors are handled and their mathematical and pedagogical quality. Thus, "teachers with weaker mathematical knowledge often addressed errors in conceptually unproductive ways, sometimes leaving errors unsolved. Often these teachers reverted to greater focus on procedures, which tended to be more closely aligned with their personal ways of understanding mathematics" (Bray, 2011, p. 31). Besides mathematical knowledge and beliefs, a third aspect identified in the study influencing teachers' error-handling practices in class discussions is their knowledge of routines and strategies to generate and facilitate discussions among students. Teachers who know some of these strategies are more likely to effectively identify the source of the errors and guide the class into a productive discussion. In contrast, teachers without this kind of knowledge tended to maintain traditional interaction patterns, including having the main control of the conversation and providing limited opportunities for students to engage in discussions and analysis of each other's ideas.

Besides the impact of teachers' knowledge in their analyses of student errors, An and Wu (2012) postulate that analyzing student errors can even support the growth of teachers' pedagogical content knowledge. In a study evaluating the impact of a model of assessing and analyzing students' homework on a daily basis, An and Wu (2012) found that by identifying errors and analyzing misconceptions, teachers' knowledge about student thinking improved considerably. This also had an impact on the quantity and quality of their feedback, as they were constantly monitoring and following up on students' understanding and were able to integrate the error patterns in their instructional practices and assessment. Similarly, Brodie (2014), in her study of teachers' professional learning communities, found that analyzing learner errors supported the participating teachers to reorganize their own professional knowledge, develop their reflective abilities and help them start learning from their teaching. By constantly analyzing common student errors in small groups over a period of three years, teachers enhanced their understanding of key concepts about learner errors; they started to see them as reasonable for the learner, as an opportunity to access student's thinking and as a starting point for further learning.

Moreover, the results of Brodie's (2014) study also provide evidence for the positive effect of teacher training on the development of their ability to analyze student errors. In a similar way, but using an intervention with different characteristics, Heinze and Reiss (2007) found that teacher training can enhance the way teachers handle errors in their classrooms. In a quasi-experimental study with students from 29 German classrooms and their teachers, they investigated the impact of teacher training on the role of errors in the teaching and learning process. They found that it had positive effects on teachers' behavior in the classroom to the extent that students realized that teachers were more tolerant and accepting of error situations and that they used them in benefit of students' learning. Results also showed that students' performance in a mathematical test improved significantly better in the group whose teachers had participated in the error training. Likewise, Holmes et al. (2013) found that teachers who participated in a three-day-workshop aimed at developing their skills to identifying and responding to student misconceptions improved their ability to discern between students' mistakes and misconceptions up to a 43% at the end of the training.

In her study with preservice secondary school teachers, Heinrichs (2015) found that even a short university seminar sequence can have an effect on the development of the diagnostic competence in error situations. Participants were 138 secondary preservice teachers from four German universities who took part in the four-lesson intervention. Analyzing preservice teachers' responses to the pre- and post-test, Heinrichs (2015) found that preservice teachers' competence was influenced in the two components under study, namely their competence to hypothesize about causes of student errors and their preferences when dealing with errors. Whereas the competence to hypothesize about causes of student errors increased significantly after the intervention, slightly more preservice teachers showed a preference towards constructivist approaches to dealing with the errors after participating in the seminar sequence. It was also concluded that the study provided evidence showing that preservice teachers' diagnostic competence in error situations "can be fostered during university teacher education and that... it seems promising to implement courses on diagnostic competence into the regular teacher education programs as well as add courses in which students' errors are analyzed more closely" (Heinrichs & Kaiser, 2018, p. 92). In the same line, Cooper (2009, p. 360) argued that "more in depth learning experiences, focused on children's thinking, during mathematics methods courses could lead to a better understanding and provide preservice teachers with a better framework for identifying important instructional strategies as they begin their first years of teaching".

1.3.3 Model of Teachers' Diagnostic Competence in Error Situations

Considering a number of different approaches from the literature that characterize the diagnostic process in teaching situations, Heinrichs (2015) developed a three-phase-model of teachers' diagnostic competence in error situations, which is used as a foundation for the present study.

The model takes into account some of the approaches representing the diagnostic competence as a process model (Heinrichs & Kaiser, 2018). As explained in section 1.2.2., these models describe and analyze teachers' thinking and decision making when they identify a discrepancy between their expectations and the actual situation in class (Peterson & Clark, 1978; Shavelson, 1983; Shavelson & Stern, 1981). Of especial interest for the model were the studies focusing on teachers' decisions made during classroom interactions and their proposals regarding the factors influencing teachers' thought, judgments and decisions when student behavior deviates from what was expected and planned. However, these models do not provide information about the characteristics of the decisions and whether they are based on a diagnosis.

For the development of the model, Heinrichs (2015) also considered the contributions of Rheinberg (1978), Reisman (1982) and Klug et al. (2013). Rheinberg (1978) developed a model that can be applied to less systematic and more spontaneous diagnostic processes. It consists of six steps, starting with student behavior and its perception by the teacher. Then, the teacher makes assumptions about the underlying causes for the behavior, which affects the teacher's reaction, which is then received by the student either implicit or explicitly and, finally, produces an effect on student behavior. Similarly, the diagnostic teaching cycle suggested by Reisman (1982) consisted of five phases that included identifying the student's strengths and weaknesses, hypothesizing about causes and creating and implementing remedial strategies. A special characteristic of Reisman's model is its cyclic nature, as the cycle needs to be started over again after evaluating the impact of the remedial strategies. More recently, Klug et al. (2013) also developed a cyclical model that refers to teachers' diagnostic processes of students' learning. It consists of three phases: the preactional, an actional and a postactional phase. In the preactional phase, teachers decide what they want to diagnose and plan how they will achieve those goals. In the actional phase, teachers systematically gather information, interpret it and arrive at a concluding diagnosis. In the postactional phase, a pedagogical plan derived from the diagnosis is implemented. This last phase is then closely connected to the first phase of the subsequent diagnostic cycle.

Regarding models focusing specifically on teachers' diagnostic decisions in error situations, Heinrichs (2015) referred, among others, to the model of Cooper (2009). This three-step-model formed the theoretical base for a seminar sequence aimed at fostering preservice teachers' ability to analyze student's work to make instructional decisions. The steps for analyzing errors in students' work consisted of the identification of the error, the search for possible causes and thinking about appropriate instructional strategies. In a similar way, Cox (1975) suggested that the identification of an error should be followed by the detection of error patterns and the selection of remedial strategies. Likewise, three phases are to be recognized in the work of Beck (2008), who suggests that in diagnostic situations, teachers identify a discrepancy between expectancies and learning behavior, implement some diagnostic strategies to analyze and understand the learning situation, so pedagogical strategies that foster student learning can be put in place.

Heinrichs (2015; Heinrichs & Kaiser, 2018) developed a model that identifies the three essential steps in the diagnostic process in error situations, which has strong connections to the noticing skills distinguished by Blömeke et al. (2015) and within the TEDS-FU project (Kaiser et al., 2015). In the first phase, the error has to be *perceived* and identified as an incorrect solution. This is an absolutely necessary step so that further actions can be taken. In the second phase, teachers have to develop hypotheses about possible causes for the error. This phase is considered by Heinrichs and Kaiser (2018) as the central element of the model because of the key role of the causes of errors for promoting further learning. In fact, this phase is present in all models of diagnostic processes they analyzed. The third phase consists of dealing with the error. Based on their hypotheses, teachers have to decide what approach they will take to support the student in overcoming their misconception and fostering learning.

Besides Heinrichs' (2015) study, other studies have used similar steps or models to characterize the diagnostic process in error situations. Peng and Luo (2009) identified four phrases or types of error analysis: identify the mathematical error, interpret the underlying rationale of the error, evaluate the student level of performance and remediate by applying a teaching strategy to overcome the error. In a study that developed an online error analysis tool for preservice teachers to support the development of their pedagogical content knowledge, McGuire (2013) designed an error analysis problem structure that included three separate but related levels. First, the student's error pattern or misconception had to be identified; then, preservice teachers were asked to "think like a student" by answering some similar problems applying the same error pattern; finally, remediation strategies had to be described. Although some level of interpretation is necessary by thinking like a student, the search for the underlying causes of the error is not explicitly included in this model. The work of Brodie (2014) did explicitly include an interpretation phase. Synthesizing various terms used in the literature referring to the analytic processes that teachers go through when they find student errors, Brodie (2014) identified three phases that in real situations occur simultaneously. As in other models, teachers start by identifying, attending to and showing an interest in the learner's error. After this, teachers interpret the error and evaluate it, for which they need both subject matter knowledge and pedagogical content knowledge. Finally, teachers decide how to handle the situation and how to engage with the error in a productive way.

The relation of these models with the one developed by Heinrichs (2015) is clear. As the present study uses this latter model, in the following each of the phases will be described in more detail, based on Heinrichs' (2015) description of the phases.

1.3.3.1   First Phase: Perceiving the Error

To start the diagnostic process in an error situation, teachers have to perceive a deviation of the circumstances from their expectations. Shavelson and Stern (1981) suggest that to cope with the large amounts of information about students, teachers select pieces of information to make their judgments. To do this, teachers constantly monitor the situation during teaching and decision making is called for when actual behavior does not occur as expected or planned. Similarly, the early models of diagnostic teaching processes of Rheinberg (1978) and Reisman (1982) include initial phases of perception or identification of student behavior. Referring specifically to errors, it is widely accepted that the first step towards working with the error in a way that fosters learning is identifying the error or error pattern (Brodie, 2014; Cooper, 2009; Cox, 1975; Riccomini, 2005; Türling et al., 2012;

Wuttke & Seifried, 2013). Furthermore, Seifried and Wuttke (2010b) state that a pedagogical response to an error is only possible if the teacher recognizes the error.

Within the noticing framework, teachers' processes of perceiving and selectively attending to relevant situations in classroom settings have also been closely investigated. Star and Strickland (2008) focus their conceptualization of noticing only on the identification of salient aspects of classroom situations. They focus on "what preservice teachers attend to -what catches their attention, and what they miss- when they view a classroom lesson" (Star & Strickland, 2008, p. 111). They also suggest that it is only possible to interpret or try to make sense of what has been noticed or perceived. Although in the noticing framework of van Es and Sherin (2002) other components are included, identifying what is noteworthy in a teaching situation is regarded as the initial key process. They state that "teachers cannot possibly respond to all that is happening in any given moment. Instead, teachers must select what they will attend and respond to throughout a lesson" (van Es & Sherin, 2002, p. 573). They regard this process as relevant for teaching because it is not possible to plan lessons completely in advance and pedagogical decisions have to be made during instruction. Similarly, attending to students' strategies is the first of the three interrelated skills included by Jacobs et al. (2010) in their conceptualization of noticing. They focus on teachers' attention to the details in the mathematical strategies students use, as they provide valuable insights into students' understandings.

The follow-up studies of the international TEDS-M study have conceptualized teachers' noticing with an emphasis on classroom situations. This model is oriented towards investigating teachers' situation-specific skills and organizes noticing into three sub-facets, namely perception, interpretation and decision-making (Yang, Kaiser, König & Blömeke, 2020). Thus, teachers' professional noticing is also initiated by the perception of particular events in teaching situations. Teachers can attend to important events from a mathematical-related or a classroom-management perspective (Blömeke et al., 2015).

1.3.3.2 Second Phase: Hypothesizing about Causes of Students' Errors

After identifying a mathematical error in a student's work or during learning situations, teachers have to investigate its causes. This interpretation of students' understanding and the development of hypotheses about causes of the incorrectness in students' thinking is a central feature of the diagnostic process (Heinrichs, 2015). Probably because of its key role, this phase can be found in most models of diagnostic processes. For instance, Jäger (1999, 2006, 2010) includes a second phase in his model of diagnostic processes in which the question to be investigated is formulated as one or more hypotheses. Later in the process, methods to collect and interpret data are used so the hypotheses can be verified or rejected. Similarly, the diagnostic models of Rheinberg (1978) and Reisman (1982) make explicit a phase in which teachers make assumptions or develop hypotheses about students' behavior. Also, Klug et al.'s (2013) cyclical model of teachers' diagnostic processes of student learning includes "making a prediction about a student's development and possible underlying learning difficulties" (Klug et al., 2013, p. 39) in the actional phase, which is the second and central dimension of their model. Similarly, the diagnostic process model developed by Wildgans-Lang et al. (2020) considers producing a hypothesis about possible sources of the identified error or learning difficulty.

Referring specifically to error analysis as a diagnostic instrument, Scherer and Moser Opitz (2012), in their guidelines for mathematics instruction at the primary school level, highlight the relevance of the hypotheses that teachers make about students' mathematical procedures and the causes of their errors as the foundation for a diagnosis that focuses on the learning process. Götze et al. (2019) emphasize that diagnosing implies not only detecting an error or an error pattern but also identifying their plausible causes. Likewise, Ashlock (2010) stresses out that when checking students' work, "we need to examine each student's paper diagnostically – looking for patterns, hypothesizing possible causes, and verifying our ideas" (p. 15). Also in the conceptualization of *professional error competence* by Seifried et al. (2015), the identification of potential causes of an error plays a key role. They state that to promote students learning from errors, besides beliefs on the potential of student errors for learning and the availability of strategies for handling the errors, "teachers need domain-specific content knowledge as well as knowledge about common student errors and potential causes for student errors" (Seifried et al., 2015, p. 178).

From a different approach, based on observing and analyzing primary teachers' practice, Ball et al. (2008) also emphasize the relevance of finding the causes of students' errors. They state that teachers have to be able to conduct mathematical error analysis fluently and rapidly, in the midst of complex instructional situations. Moreover, they suggest that to be effective, teachers need to achieve a detailed understanding of the difficulties students encounter and this "entails analyzing the source of the error" (Ball et al., 2005, p. 17), for which teachers need "to have a good hypothesis about what might be causing the error" (Ball et al. 2005, p. 18).

By suggesting that teachers need to make hypotheses about various possible causes for an error, it is implied that a particular error can have several different causes. Put in a different way, when developing hypotheses about causes of an error, each error can be assigned into different categories (Scherer & Moser Opitz, 2012). Referring to the relevance of diagnosis for making instructional decisions, Borasi (1994) points out that different types of errors would lead to different acti-vities for dealing with them in mathematics classrooms. Peng and Luo (2009) also suggest categories for the nature of mathematical errors, namely mathemati-cal, logical, strategic and psychological errors. From these and other studies, like the ones described in section 1.3.1 under the conceptualization of errors, it can be deduced that errors can be attributed to general causes or to causes that can be described as specific for that particular error. Thus, teachers can either make gene-ral hypotheses about the causes for an error, such as attributing it to psychological deficits, difficulties understanding the task or lack of subject knowledge or they can develop hypotheses that are specific for the error, considering its particular characteristics (Heinrichs & Kaiser, 2018).

In the present study, it is considered that for the successful performance of this second phase of the diagnostic process in error situations, a special compe-tence is required. Heinrichs (2015) names this competence as the *competence to hypothesize about causes of students' errors* and conceptualizes it as:

> the ability to find different hypotheses about causes for one specific error and especially being able to name causes for the specific error and not only the general reasons for an error to occur. Additionally, people with a high level of this competence are able to identify plausible and implausible causes of an error. (Heinrichs & Kaiser, 2018, p. 85)

This definition entails that teachers need to be able to take a students' perspective when analyzing the error and thinking about causes. However, a single perspective is not enough; teachers need to identify multiple sources for the error. Moreover, the search for causes is especially useful when specific causes, related to cha-racteristics of the particular error, are identified. To do this, general knowledge about teaching and learning and mathematical pedagogical knowledge is of great relevance; even more, knowledge about the role of errors in the learning process and about errors on specific mathematics topics are useful (Heinrichs, 2015).

1.3.3.3 Third Phase: Preferences for Dealing with Students' Errors

After finding causes for students' errors, teachers have to react and make appro-priate pedagogical decisions. This phase is present in many, but not in all, models of diagnostic processes. For instance, in the diagnostic teaching cycle of Reisman (1982), after the hypothesizing of reasons for learners' difficulties and strengths,

phases of "formulating behavioral objectives to serve as a structure for the remediation of weaknesses or the enrichment of strengths" and of "creating and trying
corrective remedial procedures" (Reisman, 1982, p. 5) are included. In other
models, instructional reactions are not an obligated final stage of the process,
as in the model of Jäger (2006). He suggests that in the final phase, the obtained
diagnostic is communicated and can also lead to the implementation of further
educational provision for learners. Hence, in cases in which the aim of the diagnostic process is providing a psychological report, the process ends with the
communication of the results, whereas when the goal is remediation, an intervention is designed and then diagnostically accompanied (Jäger, 1999, 2006).
Similarly, the model of Wildgans-Lang et al. (2020) for diagnostic activities
finalizes with the communication of the diagnosis to students, parents or other
professionals. Thus, it does not include an instructional reaction on the side of
the teacher.

By contrast, Edelenbos and Kubanek-German (2004) attribute a significant
role to pedagogical responses in their conceptualization of diagnostic competence.
They argue that provision of appropriate support, in which teachers scaffold learning, should follow as a response to the diagnosis. Likewise, the conceptualization
of situation-based diagnostic competence by Hoth et al. (2016) emphasizes not
only the findings made in the diagnostic process but also the decisions to modify
instruction at the micro-level during teaching. Another model including this third
phase of making pedagogical decisions is the one developed by Klug et al. (2013).
They consider that the aim of a diagnosis is teachers' understanding of students'
learning behavior so that they can support students individually and adapt instruction to their needs. Therefore, in the third phase of their model, the postactional
phase, teachers implement a pedagogical plan that is derived from the diagnosis.
This plan may include providing feedback to students and parents, developing
individualized pedagogical strategies or adapting whole-class instruction.

Despite the models considering that in some situations the diagnostic process
can end in the understanding or judgment of student learning or its communication, models that refer specifically to error situations agree in the need to take
actions oriented towards the modification or remediation of the error. In learning
situations, errors are to be used as learning opportunities and as the starting point
for building further knowledge and skills (Borasi, 1994; Cox, 1975; Heinrichs,
2015; Prediger & Wittmann, 2009; Rach et al., 2013; Radatz, 1980; Santagata,
2005; Schleppenbach, Flevares, Sims & Perry, 2007; Seifried et al., 2015; Smith
et al., 1993; Son, 2013). Götze et al. (2019) point out that to promote learning,
errors need to be identified and discussed. Even more, they argue that without

fostering activities, a diagnosis remains useless. Unfortunately, despite the existence of theoretical and empirical evidence of the relevance of teachers' decisions in error situations, only very few statements on how to deal with students' errors are available (Reiss & Hammer, 2013; Seifried & Wuttke, 2010a).

This involves a significant obstacle in making judgments about the appropriateness of preservice teachers' suggestions on how to deal with the errors included in the present study. Therefore, responses will be examined in terms of their characteristics in three aspects: their focus towards promoting conceptual or procedural understanding of mathematics, the individualization of the proposal to the particular characteristics of the error situation and the allusion of an active-learner or a teacher-directed approach. These three aspects are then used to code the open items and thus to classify preservice teachers' preferences for dealing with the errors by the identification of different types of approaches with the aid of latent class analysis, as explained in section 3.3.2.2.

Since the strategies preservice teachers choose to deal with errors is more a preference for different teaching approaches, this component of diagnostic competence in error situations is associated in the present study less to the cognitive part of the concept of competence and more to the convictions or orientations of teachers. This relates to the concept of orientations introduced by Schoenfeld (2011), which refers to the values, beliefs and preferences that guide and shape decisions individuals make in different contexts. Because orientations are very closely linked to the concept of beliefs, the relevance of non-cognitive features in the context of preservice teachers' preferences for dealing with errors becomes clear. In what follows, the three aspects used to examine preservice teachers' preferences for dealing with student errors in the present study will be described.

Conceptual and Procedural Understanding
Previous studies in mathematics education have distinguished between procedural and conceptual knowledge and their role in the development of mathematical competence. Actually, "there is a long-standing and ongoing debate about the relations between [these] two types of knowledge" (Rittle-Johnson, Schneider & Star, 2015, p. 588).

Conceptual knowledge has been defined as the "explicit or implicit understanding of the principles that govern a domain and the interrelations between pieces of knowledge in a domain" (Rittle-Johnson & Alibali, 1999, p. 175). This knowledge is flexible and independent of specific problems or situations because it implies abstract concepts, relations and general principles (Lenz, Dreher, Holzäpfel & Wittmann, 2019; Rittle-Johnson, Siegler, Alibali, 2001). According to Zahner, Velazquez, Moschkovich, Vahey and Lara-Meloy (2012), one key goal of mathematical education is the development of students' conceptual understanding and, to achieve this, teaching has to explicitly focus on mathematical concepts and support students in making connections among them. Procedural knowledge is defined as "action sequences for solving problems" (Rittle-Johnson & Alibali, 1999, p. 175). In other words, is knowledge about the procedure or sequence of steps to do something or achieve a goal (Lenz et al., 2019; Rittle-Johnson et al., 2015; Rittle-Johnson et al., 2001). Three decades ago, Porter (1989) reported that mathematical instruction in elementary schools failed to give sufficient attention and time to problem solving and conceptual understanding, allocating more than 70% of teaching time to skill development. More recently, analyses of secondary school curricula and textbooks show that, although opportunities to develop both conceptual understanding and procedural fluency are provided, the majority of tasks focuses on procedural knowledge (Lenz, Holzäpfel & Wittmann, 2019; Son & Senk, 2010). Ashlock (2010) highlights the importance of developing both types of knowledge. He argues that "both conceptual learning and procedural learning are necessary, but procedural learning needs to be tied to conceptual learning and to real-life applications" (Ashlock, 2010, p. 7).

Although there is consensus that both types of knowledge are needed to be competent in mathematics, the relationship between them has been controversial (Rittle-Johnson et al., 2015). The positive effects of conceptual knowledge on the development of procedural knowledge are widely accepted. However, it remains unclear if the same occurs in the opposite direction. According to Rittle-Johnson and Koedinger (2009), rather than one type of knowledge preceding the other, conceptual knowledge and procedural knowledge do not develop independently, they influence each other. They propose an iterative model, in which the development of one type of knowledge leads to an increase in the other, which then elicits an increase of knowledge in the former (Rittle-Johnson et al., 2001; Rittle-Johnson & Koedinger, 2009). In an article examining these relations, Rittle-Johnson and colleagues (2015) concluded that:

> evidence indicates that the relations between conceptual and procedural knowledge are often bidirectional, with improvements in procedural knowledge often supporting improvements in conceptual knowledge as well as vice versa. It is not a one-way street from conceptual knowledge to procedural knowledge; the belief that procedural knowledge does not support conceptual knowledge is a myth. (Rittle-Johnson et al., 2015, p. 594)

Despite the need to develop both types of knowledge, Ashlock (2010) argues that when interpreting student performance and their error patterns, a distinction needs to be made between cases in which students lack conceptual understanding and when they do not know the correct procedures. This is especially relevant in the area of arithmetic and has implications for the selection of instructional responses and strategies.

Son and Sinclair (2010) have used the distinction between these two types of knowledge to investigate preservice teachers' interpretation and response to a student's error in geometry. They found that a large number of preservice teachers suggested responses based on procedural aspects, even when they had attributed the error to conceptual aspects. In a later study, Son (2013) found that preservice teachers did provide a response corresponding to their interpretation of the error as a conceptual-based or a procedure-based error. However, results also indicated a clear tendency towards attributing the error to a lack of procedural knowledge despite the fact that the student's error suggested a lack of conceptual knowledge. Likewise, Chauraya and Mashingaidze (2017) investigated in-service secondary school mathematics teachers' interpretation of students' errors in algebra. They examined teachers' understanding of the students' reasoning behind some common errors in algebra. Besides showing teachers difficulties in explaining students' errors in algebra, results provided evidence that, when correct explanations were given, they focused mostly on procedures, without linking the errors to conceptual aspects. A similar result was found by Cooper (2009) in her study with preservice teachers. She found that most of the strategies suggested by preservice teachers to deal with student errors were oriented to 're-teaching' the topic with a focus on procedures or rules.

In a case study, Bray (2011) investigated the relation between third-grade teachers' beliefs and knowledge and their practices in dealing with students' errors. One of the dimensions she identified that varied across teachers was the extent to which they foster conceptual understanding of mathematics. She found that teachers' beliefs played a key role in their intentions to focus on conceptual understanding, in that "teachers who emphasized mathematics concepts were more likely to believe that understanding mathematics concepts is more powerful and more generative than remembering mathematics procedures" (Bray, 2011, p. 29). Her results also suggest a close relation between teachers' knowledge and the likelihood that their responses to students' errors in class would promote conceptual understanding. Stronger knowledge supported teachers' ability to anticipate errors and emphasize mathematics concepts in their responses to errors, whereas weaker knowledge led to difficulties in unpacking the underlying reasons for errors. She states that "teachers with weaker mathematical knowledge often addressed errors in conceptually unproductive ways, sometimes leaving errors unresolved. Often, these teachers reverted to greater focus on procedures, which tended to be more closely aligned with their personal ways of understanding mathematics" (Bray, 2011, p. 31).

Besides focusing to a greater or lesser extent on promoting students' conceptual or procedural understanding of mathematics, teaching strategies can vary in several other aspects. For instance, a range of specificity can be observed in teaching strategies used in error situations. This differentiation will be described next.

Targeted Strategies
A key goal in identifying and interpreting students' errors is to deal with error situations in a way that fosters further learning. Instruction needs to be focused on student learning by individualizing strategies and by starting further learning with what the student already knows and build on that knowledge (Ashlock, 2010). Moreover, teachers' understanding of student thinking should lead to more targeted instruction in which tasks, questions or other strategies are designed to address underlying foundational gaps in understanding (Van de Walle, Lovin, Karp & Bay-Williams, 2014). Nevertheless, aligning instructional responses to students' confusions and to the particular concepts to be taught has been acknowledged as a challenging task (Choy, Thomas & Yoon, 2017).

In fact, in a study examining pre- and in-service teachers' diagnostic competence in mathematics teaching situations, Biza, Nardi and Zachariades (2018) found great variation in the extent to which responses targeted the specific situation. A wide range of specificity was observed in the responses, from approaches that strongly accounted for characteristics of the incident presented in the task to others being only slightly related to the incident. Similarly, Sánchez-Matamoros, Fernández and Llinares (2019) explored the relationships between preservice secondary teachers' skills of attending to key aspects in students' mathematical work, interpreting their understanding and making instructional decisions. They concluded that "deciding how to respond on the basis of students' mathematical understanding was the most challenging skill" (Sánchez-Matamoros et al., 2019, p. 96). The instructional actions provided by preservice teachers rarely focused on proposing conceptual actions concerned with fostering understanding; most of them suggested strategies that were categorized as 'general actions', like explaining the content all over again, or as 'procedural actions', oriented towards practicing formulas or procedures. Moreover, they found that preservice teachers who could identify the connections between key mathematical elements in students' work and students' understanding were more likely to plan instructional activities that foster students' comprehension of mathematical ideas. They argue that these results suggest a connection between preservice teachers' identification and interpretation skills and their decision-making skill and emphasize the link between noticing skills and teachers' professional knowledge.

In line with these findings, in their study investigating preservice teachers' interpretation and response to a student's error in geometry, Son and Sinclair (2010) found that only 11 out of the 54 responses were specifically tailored to address the student's difficulty with the task. They uncovered that a great majority of preservice teachers offered general strategies that might cover various difficulties. A closer analysis of the responses permitted the identification of three types of strategies, which did not directly address the student's error. The first involved the generalization of properties, in that preservice teachers would invoke general properties of the topic by 'showing', 'telling' or 'talking about' properties that were more general than those needed to understand the specific task. The second type of strategy consisted of returning to the basics, going too far back to the first principles, even when the student may not have needed to go so far back to address her error. They called the third type of general strategy 'Plato-and-the-slave-boy' approach. It involves the assumption that the student actually did know the concept or understand the topic but had simply forgotten. Thus, the response consisted of reminding the student of the 'forgotten contents' or asking her to remember them. In a later study, Son (2013) also included the analysis of the level of use of the student error in preservice teachers' interpretations and responses to an error in the topic of ratio and proportion in similar rectangles. She classified the extent to which the student's error was utilized for the suggested responses into three levels: active use, intermediate use and rare use. An active use of the error was observed when preservice teachers used the student's strategy as a central tool of their pedagogical response and provided opportunities to discuss and investigate why the student's method was not correct. When the student's error was used but did not play a central role, it was considered an intermediate use. In these cases, the student error was briefly addressed as a trigger to further actions but not really discussed. Responses in the third level rarely used the student error, they tended to make a statement about the incorrectness of the method or not to mention it at all. The main and only focus in this level was on the correct procedure.

In addition to variations in their focus on either conceptual or procedural understanding and the extent to which their design is specifically tailored to address students' learning needs, teaching practices can show differences in their underlying approaches to teaching. Particularly, differences can be found in teachers' preferences regarding student-teacher interactions in the classroom and characteristics of the role each of them plays in the learning process. These differences will be described in what follows.

Teacher Direction and Active Learner Approaches
Fundamental differences can be found in the roles that students and teachers play in contrasting approaches to addressing errors in classroom situations. On the one side, there are teaching approaches in which most of the activities are centered on the teacher, who is most of the time explaining, showing, telling and delivering information to students. On the other side, there are approaches in which students play the central role and the teacher provides guidance and opportunities for students to think, do, experiment and discuss. These two approaches can be understood as ends of a continuum, with teachers sometimes combining both but usually showing preferences in one direction. This differentiation is explained by Helmke (2017) using the terms 'Instruktion –Konstruktion'. He claims that dichotomizing both elements is meaningless, as both are relevant for school learning. Aspects such as motivation, guidance or explanations involve both teachers and students.

A similar differentiation was used in the international comparative study of mathematics teacher education 'Mathematics Teaching for the Twenty first Century' (MT21), which distinguished between a traditional and directive-instruction approach and an active and self-directed learning approach (Müller, Felbrich & Blömeke, 2008). In the former approach, the teaching and learning process is subdivided into small steps and strongly controlled and led by the teacher; drill and practice, receptive student learning and lack of interactive aspects are at the core of the process. The latter approach is characterized by student self-direction, learning by discovering and cooperative learning strategies. In this approach, students are actively engaged in numerous mathematics activities. Particular characteristics for dealing with error situations from both approaches were specified within the study. Whereas in traditional approaches, errors are avoided, ignored or quickly corrected to continue with the lesson, in active learning approaches, errors are seen as opportunities for further learning, they are intensively discussed by students and teachers' questions are aimed at promoting argumentation of the mathematical ideas involved.

In a smaller study examining preservice teachers' analyses of students' work, Cooper (2009) made a similar distinction. Besides determining if they were able to identify the computational error patterns and determine reasonable underlying misconceptions, the instructional strategies recommended by preservice teachers were qualitatively analyzed. Among other themes, responses were categorized into strategies that would deal with students' errors from a 'teacher-directed instruction' approach and those that would take a 'student-teacher interaction' approach. The first was characterized by responses that focused on the teacher showing how to complete a task or modeling a procedure. In the second approach, preservice teachers' responses suggested a stronger involvement of students, in that they would be asked to explain their mathematical reasoning and make connections between mathematical ideas.

An analogous distinction can be found in the work of Son and Crespo (2009), who examined preservice teachers' reasoning and responses to a student's non-traditional strategy for dividing fractions. To analyze preservice teachers' responses, they developed two categories to approach mathematics teaching, namely 'student-focused' and 'teacher-focused'. Student-focused responses were those suggesting that students should be provided with opportunities to explain and justify their division strategy and that teachers should be guiding the discussion towards the understanding and exploration of the method's correctness and efficiency. On the contrary, teacher-focused responses suggested that teachers would determine if the method is or not correct and would show, tell and explain the topic to students. Likewise, in another study analyzing preservice teachers' interpretation and responses to students' geometric errors, Son and Sinclair (2010) found that responses addressed student errors either by showing and telling or by giving and asking. In the first form, teachers delivered verbal or nonverbal information to the student, for instance, by explaining or modeling a procedure, showing an image or telling a fact. In the second form, the student was requested to provide some information (verbal or nonverbal), to do something, manipulate some material or produce some representation. Results showed that most preservice teachers provided responses in the first form, revealing a preference for delivering information. It was concluded that "despite current reform efforts aiming to change teaching practice, the notion of teaching as telling (showing, explaining) rather than facilitating (listening, interpreting) still pervades among the preservice teachers in this study" (Son & Sinclair, 2010, p. 41).

In a later study concerned with preservice teachers' interpretation and responses to students' errors, Son (2013) used a similar distinction to evaluate the nature of teacher responses. She differentiated between a 'show-tell' approach and a 'give-ask' approach. Like in the previous study, results indicated that most preservice teachers showed a preference for giving information rather than providing opportunities for students to discuss and elaborate on their understandings.

The distinction between these two ways of addressing students' errors was also used in the study design of Türling et al. (2012). In a study of teachers' professional error competence, video vignettes presenting short classroom error situations were used as prompts for an interview in which teachers' competence to identify and respond to errors were evaluated. Based on the errors teachers identified in the first video, the causes they attribute the error to and how they would handle the error, the second video vignette was selected (by the interviewer) from four possible follow-up sequences. These sequences varied regarding two criteria, namely the possibility to work with a single student or the whole group and the extent to which teachers would give students clues for correctly solving the task. The latter criterion can be related to the distinction between teacher-direction and active-learner approaches. The provision of strong and explicit hints can be attributed to a teacher-direction approach and less explicit clues, in which more initiative from the student is expected can be related to an active-learner approach.

Similarly, Heinrichs (2015) differentiated between instructivist and constructivist approaches in her study analyzing preservice secondary teachers' diagnostic competence in error situations. The results suggested that preservice teachers' individual preferences when dealing with student errors "could be differentiated into three classes of teachers, who either preferred a constructivist or an instructivist approach or used both approaches flexibly" (Heinrichs & Kaiser, 2018, p. 90).

The present study differentiates between strategies to handling students' errors that are based on a teacher-directed approach and those that provide evidence of taking an active learner approach. The former is recognized on instructional strategies focusing mainly on teachers' activity, as they are who are delivering knowledge, explaining procedures or giving examples of mathematical ideas. On the contrary, the active learner approach focuses on what the student is doing, on the activities, tasks, examinations, discussions or reasoning he is going through to develop his mathematical abilities.

Research Question and Hypotheses

2

The present study aims at examining the development of preservice primary teachers' diagnostic competence in error situations. It takes as a framework the concepts of teachers' professional competence and diagnostic competence and the role of students' errors in mathematics teaching and learning discussed in Chapter 1. In particular, the design of the study is based on Heinrichs (2015) definition of teachers' diagnostic competence in error situations, published in English as

> The competence that is necessary to come to implicit judgements based on formative assessment in teaching situations by using informal or semi-formal methods. The goal of this process is to adapt behavior in the teaching situation by reacting to the student's error in order to help the student to overcome his/her misconception. (Heinrichs & Kaiser, 2018, p. 81)

As has been argued in Chapter 1, in order to provide adequate learning opportunities for all students in diverse classrooms, teachers need to have the ability to understand students' thinking. Furthermore, students' errors provide teachers with a rich source of information about students' reasoning and strategies. Despite the crucial role that teachers' diagnostic competence plays in the learning process, it has received limited attention, especially in Spanish-speaking countries. Often, this aspect of teachers' professional competences is not even considered in university programs. This study aims at contributing to this situation by examining how preservice primary teachers' diagnostic competence in error situations can be fostered within initial teacher education.

The model of diagnostic competence in error situations developed by Heinrichs (2015) and published in English by Heinrichs and Kaiser (2018) is taken

M. Larrain Jory, *Preservice Primary Teachers' Diagnostic Competences in Mathematics*, Perspektiven der Mathematikdidaktik, https://doi.org/10.1007/978-3-658-33824-4_2

as a base to structure both the design of the university seminar sequence and the instruments that permitted the evaluation of the development of the competence. The model was also translated into an error analysis cycle that was used by preservice teachers as a facilitating tool for diagnostic thinking.

The study also took a situated perspective, in that strategies were used to make the analysis of error situations as close to reality as possible. Samples of students' work and short videos were used to achieve this. However, it is not the study's aim to come up with statements about the actual behavior of preservice teachers in diagnostic situations in the classroom, but to describe their preferences and their thinking about students' errors in situations presented in written and video formats. This may, of course, differ to some extent from what they would actually think or do in a real classroom context. These possible differences are also not to be covered by the present study.

Taking all this into consideration, this study seeks to answer the following research question:

To what extent is it possible to promote preservice primary school teachers' diagnostic competence in error situations within initial teacher education?

To deal with this question, the statements were divided into three hypotheses. The prove or disprove of these hypotheses based on the collected data will provide information leading to answering the research question. The sample size allows exploring the correlations between preservice teachers' diagnostic competence and other background variables. However, the results are influenced by this particular sample and cannot be generalized. In other words, the present study does not claim general validity.

The three hypotheses covered aspects of the conceptualization of the diagnostic competence in error situations and features related to it, the development of the competence within initial teacher education and the aspects that may have an influence on this development during the university seminar sequence. To scrutinize these hypotheses, they were particularized into various sub-hypotheses so they could be verified using statistical tests.

First, in order to better understand the construct of diagnostic competence in error situations and aspects related to it, this study seeks to examine the following hypothesis:

Hypothesis 1: *Preservice primary school teachers' diagnostic competence in error situations is related to their beliefs, knowledge and opportunities to learn.*

It is possible to assume that preservice teachers' ability to understand students' thinking, to interpret the reasoning underlying their mathematical errors and to adapt teaching strategies accordingly, might be related to various features of their teaching competence, including their beliefs about the subject and about teaching, their professional knowledge and their practical experience.

One important feature is preservice primary teachers' beliefs. As errors play an important role in the learning process under a constructivist paradigm, it can be assumed that constructivist beliefs about teaching and learning would lead to a greater awareness of the relevance of errors in promoting students' learning. Similarly, someone with beliefs about the nature of mathematics as a process of inquiry may be more eager to interpret students errors with flexibility than someone who beliefs mathematics is a set of rules and procedures that need to be learned and applied, who may see errors as something to be ignored and who would simply restart the learning of the rules and procedures involved.

Another aspect that may be related to preservice teachers' diagnostic competence is their professional knowledge and the learning opportunities they have experienced. Mathematical knowledge for teaching may provide a strong knowledge base with which student reasoning can be interpreted and pedagogical decisions can be made. Moreover, the number of university courses on mathematics and mathematics education preservice teachers have completed during their teacher preparation is expected to have contributed to their mathematical knowledge and their pedagogical content knowledge and is therefore assumed to be related to their competence to analyze students' errors and make pedagogical decisions. Additionally, the number of completed semesters in their teacher education programs is also expected to be related to their diagnostic competence, as preservice teachers in higher semesters have had more learning opportunities. Not only mathematics or mathematics education courses may play a role, but also other courses related to educational psychology, learning theories, curriculum, evaluation and a wide variety of contents and experiences.

Furthermore, it was also considered that university entrance test scores may be related to preservice teachers' diagnostic competence. Selection test scores are usually a strong indicator of many professional competences and their development during university education; therefore, they were expected to be relevant in this case as well.

Besides beliefs and knowledge, practical experiences were considered an additional feature related to preservice teachers' diagnostic competence. The literature suggests that teachers' diagnostic competence develops over time, that teachers with more classroom experience are more skilled at identifying, interpreting and handling students' errors. Within teacher education programs, school-based placements are opportunities for preservice teachers to learn and practice the identification of errors and understanding students' reasoning and to observe and train their ability to make ongoing pedagogical decisions. In most Chilean teacher education programs, practicum has an important place. Preservice teachers participate in a sequence of school-based practicum, usually starting with a focus on observation activities, with a mostly passive role, and stepwise evolve to more active participation involving some to many whole-class teaching experiences. Evidently, completing a school practicum in which the main purpose is to observe another teacher's activity is a very different learning opportunity from those school practicums that include activities in which the preservice teacher must take responsibility for teaching during one lesson or during several lessons. Furthermore, as primary teachers in Chile are trained as generalists, learning opportunities related to the teaching of mathematics are limited and school practicums do not always include opportunities to teach mathematics. Therefore, different effects can be expected on the development of the diagnostic competence of preservice teachers who have had the experience of teaching mathematics in the context of such practices and those who have taught other subjects, but not mathematics. For these reasons, the survey differentiated between the number of completed school practicum, preservice teachers' experience teaching any subject and their experience teaching specifically mathematics in primary classrooms. Additionally, as many preservice teachers offer private tutoring and this may also be considered a rich source of learning opportunities to develop diagnostic competence, the relationship between private tutoring experience and the development of the competence is an interesting aspect to analyze.

As there are many aspects included in the statement on Hypothesis 1, it was broken down into more detailed assumptions. These sub-hypotheses are detailed below.

Hypothesis 1.1: Preservice primary school teachers' constructivist beliefs about teaching and learning mathematics and about the nature of mathematics as a process of inquiry are positively related to their diagnostic competence.

Hypothesis 1.2: Preservice primary school teachers' mathematical knowledge for teaching is positively related to their diagnostic competence.

Hypothesis 1.3: The number of mathematics and mathematics education courses completed by preservice primary school teachers' is positively related to their diagnostic competence.

Hypothesis 1.4: Preservice primary school teachers' study progress within their initial teacher education program is positively related to their diagnostic competence.

Hypothesis 1.5: University entrance test scores are positively related to preservice primary school teachers' diagnostic competence.

Hypothesis 1.6: The number of school practicum completed by preservice primary school teachers is positively related to their diagnostic competence.

Hypothesis 1.7: Preservice primary school teachers' experience teaching any subject in primary school classrooms is positively related to their diagnostic competence.

Hypothesis 1.8: Preservice primary school teachers' experience teaching mathematics in primary school classrooms is positively related to their diagnostic competence.

Hypothesis 1.9: Preservice primary school teachers' experience offering private tutoring to school students is positively related to their diagnostic competence.

Second, the central hypothesis of this study refers to the influence of a structured seminar sequence on the development of the competence:

Hypothesis 2: *It is possible to positively influence the development of preservice primary school teachers' diagnostic competence in error situations within initial teacher education by a structured seminar sequence.*

As the research question of this study refers to the influence of an intervention, in the form of a university seminar sequence, on the development of preservice teachers' diagnostic competence in error situations, this hypothesis is fundamental for the present study. It is examined based on the framework of Heinrichs' (2015) model of diagnostic competence in error situations. This means that both preservice teachers' competence to hypothesize about causes of students' errors and their preferences for dealing with error situations are studied.

It is expected that the competence to hypothesize about causes of students' errors is positively influenced by the university seminar sequence. It provides significant learning opportunities so that preservice teachers can recognize the relevance of errors as a rich source of information about students' mathematical reasoning and receive structured support on interpreting and understanding students' thinking. As the competence is measured both before and after the seminar sequence, i.e., a latent ability score is calculated for each participant at the pre-test and again at the post-test, both values can be compared.

Regarding the influence of the seminar sequence on preservice primary school teachers' preferences for dealing with students' errors, it is interesting to investigate the patterns in which preservice teachers' preferences change. Therefore, a pre- and a post-test latent class of preferences for dealing with students' errors can be estimated and changes can be described for this aspect of preservice teachers' competence. Additionally, the relationship between the changes in the competence to hypothesize about causes of students' errors and the changes in the preferences for dealing with students' errors will be examined.

The above described three aspects of this hypothesis can also be formulated in the following three sub-hypotheses:

Hypothesis 2.1: Preservice primary school teachers' competence to hypothesize about causes of students' errors can be positively influenced by a structured seminar sequence within initial teacher education.
Hypothesis 2.2: Preservice primary school teachers' preferences for dealing with students' errors can be influenced by a structured seminar sequence within initial teacher education.
Hypothesis 2.3: The changes in preservice primary school teachers' competence to hypothesize about causes of students' errors and the changes in their preferences for dealing with students' errors are related to each other.

Moreover, it is possible to assume that the influence of the university seminar sequence may differ among participants and that these differences may be explained by other factors. This leads to the third hypothesis of this study:

Hypothesis 3: *The gains on the development of preservice primary school teachers' diagnostic competence in error situations are related to their beliefs, knowledge and opportunities to learn.*

In a similar way as various factors of preservice teachers' knowledge, beliefs and opportunities to learn are assumed to be related to the observed characteristics of their diagnostic competence, those factors may also be related to the gains preservice teachers experience in the development of their diagnostic competence during the university seminar sequence. Those factors may shape the ways in that preservice primary school teachers are able to participate in the activities and the readiness they have to make connections among different pieces of knowledge and experiences. For instance, because of the great significance of errors within constructivist learning theories, preservice teachers with constructivist beliefs about mathematics teaching and learning may present a better disposition to learning about error analysis and interpreting and understanding students' thinking. Similarly, preservice teachers seeing mathematics as a dynamic subject, i.e., as a process of inquiry, maybe keener to interpret the mathematical processes and reasoning used by students from various perspectives than those who agree to a more static view of the nature of mathematics.

Another set of factors that may influence preservice teachers' gains during the university seminar sequence is related to their professional knowledge and learning opportunities. Preservice teachers with a stronger mathematical knowledge for teaching may be able to activate this knowledge during the activities and make connections between new and previous pieces of knowledge that may, in turn, contribute to a greater gain on the development of their diagnostic competence. The number of mathematics or mathematics education courses they have completed may have a similar influence by providing a knowledge base to be activated and further connected during the seminar sequence. Likewise, preservice teachers who have completed more semesters in their teacher education programs may have a wider knowledge base available for interpreting students' thinking and making decisions about how to better deal with students' errors, including knowledge from courses on the areas of educational psychology, curriculum, evaluation and others. Moreover, higher university entrance test scores, which are usually strongly related to academic and learning abilities, may be related to greater gains in the development of the diagnostic competence in error situations because of a better predisposition to learning and better learning abilities.

Additional factors that may be associated with the changes in preservice teachers' diagnostic competence in error situations are related to their practical experiences. It is possible to assume that preservice teachers who have more practical experience can appreciate better the relevance of interpreting students' errors and deciding how to deal with them than those who have not been involved in such error situations. Furthermore, more practical experiences can also increase the frequency with which preservice teachers are exposed to identifying, interpreting and dealing with students' errors on their own and observing other professionals in the process. This can also enrich their awareness about the importance of analyzing and dealing with errors, as well as their repertoire of alternatives for doing so. School practicum, in general, may contribute in this sense, but also preservice teachers' experiences leading the learning process themselves may have a particular value. Moreover, as preservice primary school teachers in Chile are usually trained as generalists, their experiences teaching specifically mathematics to primary students may have an even higher value, as it may involve situations more similar to those being included in the seminar sequence and in the assessment of this study. Lastly, one-to-one practical teaching experiences in the form of private tutoring lessons may also influence how preservice teachers interpret and how much they benefit from the seminar sequence.

Finally, it is intended that the university seminar sequence influences the development of preservice primary school teachers' diagnostic competence in error situations, so it can be presumed that the attendance to a higher number of sessions should increase their opportunities to learn about students' mathematical errors, their interpretation and handling and thus, positively influence the development of their diagnostic competence in error situations. Besides attending the sessions, active participation or preservice teachers' perceptions of their own involvement in the activities may be associated with more positive changes in their diagnostic competence. Implicit in more active participation is a more intensive immersion in the topic and also a stronger commitment to the activities and the development of the own competence.

These relationships between the gains and other factors can also be formulated into the following sub-hypotheses:

Hypothesis 3.1: The gains in the development of preservice primary school teachers' diagnostic competence in error situations are related to their beliefs about teaching and learning mathematics and about the nature of mathematics.
Hypothesis 3.2: The gains in the development of preservice primary school teachers' diagnostic competence in error situations are related to their mathematical knowledge for teaching.
Hypothesis 3.3: The gains in the development of preservice primary school teachers' diagnostic competence in error situations are related to the number of mathematics and mathematics education courses they have completed.
Hypothesis 3.4: The gains in the development of preservice primary school teachers' diagnostic competence in error situations are related to their study progress within their initial teacher education program.
Hypothesis 3.5: The gains in the development of preservice primary school teachers' diagnostic competence in error situations are related to their university entrance test scores.
Hypothesis 3.6: The gains in the development of preservice primary school teachers' diagnostic competence in error situations are related to the number of school practicum they have completed.
Hypothesis 3.7: The gains in the development of preservice primary school teachers' diagnostic competence in error situations are related to their teaching experience in primary school classrooms.
Hypothesis 3.8: The gains in the development of preservice primary school teachers' diagnostic competence in error situations are related to their experience teaching mathematics in primary school classrooms.

Hypothesis 3.9: The gains in the development of preservice primary school teachers' diagnostic competence in error situations are related to their experience giving private tutoring to school students.

Hypothesis 3.10: The gains in the development of preservice primary school teachers' diagnostic competence in error situations are related to their attendance to the seminar-sequence sessions.

Hypothesis 3.11: The gains in the development of preservice primary school teachers' diagnostic competence in error situations are related to their self-reported active participation in the seminar-sequence sessions.

The hypotheses and sub-hypotheses heretofore formulated will be explored based on the collected data and using qualitative and quantitative research methods. The methodology and methods used to do so will be described in the next chapter.

Methodological Approach of the Study

3

This chapter explicates the methods used to investigate the research question of the present study: to what extent preservice primary school teachers' diagnostic competence in error situations can be fostered within a university seminar sequence. First, it provides a description of the design of the study, including a characterization of the university seminar sequence, information about the instruments used and the testing design and a detailed description of the error tasks at the core of the instrument designed to measure preservice teachers' diagnostic competence as a latent variable. The second section provides information about the groups of preservice teachers who participated in the study and the data collection techniques. The third section explains the qualitative and quantitative methods used to analyze the data collected in the questionnaires.

3.1 Study Design

To evaluate how preservice primary teachers' diagnostic competence in error situations could be fostered, an intervention study with a pre- and post-test design was used. The intervention consisted of a university seminar sequence of four 90-minute sessions, in which preservice teachers engaged in discussions about children's thinking in error situations shown in written samples of student's work and short video clips from primary school classrooms. Prior to the seminar sequence, participants answered questionnaires that collected demographic information, data on their beliefs about the nature of mathematics and about mathematics teaching and learning, their mathematical knowledge for teaching and their diagnostic competence in error situations. After the seminar sequence,

© The Author(s), under exclusive license to Springer Fachmedien Wiesbaden GmbH, part of Springer Nature 2021
M. Larrain Jory, *Preservice Primary Teachers' Diagnostic Competences in Mathematics*, Perspektiven der Mathematikdidaktik,
https://doi.org/10.1007/978-3-658-33824-4_3

they provided information about their participation in the sessions and answered a test on diagnostic competence in error situations.

3.1.1   Description of the University Seminar Sequence

The short university seminar sequence designed for the intervention was composed of four 90-minute sessions. It was aimed at fostering the development of preservice primary teachers' diagnostic competence as used in error situations. Because for most participants it was their first approach to recognizing errors as a way to understand students' thinking, an overarching goal was to sensitize preservice teachers about the important role errors can play in the teaching and learning of mathematics.

The design of the seminar sequence was founded on the diagnostic competence model developed by Heinrichs and Kaiser (2018) described in section 1.3.3. The model provided preservice teachers with a useful tool to structure the analysis of students' mathematical errors. It was presented as an error analysis cycle with questions associated with each of its three parts that served as a guide to identify and interpret students' errors, think about a wide variety of possible causes for the errors and discuss different pedagogical approaches to deal with them and promote students' understanding. Furthermore, the structures of the seminar sequence and of each of the sessions were similar to those developed by Heinrichs (2016), adapted to take into account contextual characteristics and to meet the needs of preservice primary school teachers.

Taking into account results, evidence and suggestions from the studies presented in section 1.2.5, short videos and samples of primary students' written work were used to bring the errors into the sessions and prompt the discussions and analyses of students' thinking. To support and enrich the analyses, extracts from the literature relevant to the errors were handed to preservice teachers and included in the discussions.

It is important to highlight that it was not the aim of the seminar sequence to study deeply any particular error or to cover a wide variety of mathematical errors that may often arise in primary classrooms. The focus was on the development of preservice teachers' diagnostic competence in a more generic way. However, in order to narrow down the mathematical knowledge and mathematics-pedagogical content knowledge required to analyze the errors and to facilitate the transfer of newly acquired knowledge and abilities, all the errors used in this study (both during the seminar sequence and in the tests) belonged to the area of numeracy and operations.

The first session was aimed at sensitizing preservice teachers about the value of error analysis for improving teaching and learning and at getting to know and use the error analysis cycle with its supporting questions. The session focused on a multiplication fact error, specifically when one of the factors is zero and students' answer is the number of the other factor, i.e., n * 0 = n. This error was chosen because it is simple and appears often among students. It has also been identified and described by various authors. Padberg and Benz (2011) found that this error was the most common one leading to wrong results in written multiplication algorithms. Van de Walle et al. (2014) suggest that although it may seem procedurally easy, students can be confused when they transfer some rules they may have learned for addition. Additionally, Padberg and Benz (2011) suggest a variety of other reasons that may lead students to this misconception.

The session started with a short video clip showing students working individually on a multiplication facts worksheet. The teacher approaches a student who has recently written down five as the answer for 5 * 0 = __. Preservice teachers took notes and then briefly commented on the video with a partner. This analysis was made without any further guidance or framework, so any ideas would come out. After these ideas were commented on with the whole group, participants read and discussed an extract of an article (Larrain, 2016) presenting the role of errors that was specially intended to sensitize preservice teachers. Based on this, the error analysis cycle and guiding questions for each step were presented. Then, participants watched the video clip again and, in pairs, systematized the analysis ideas using a worksheet with the three steps of the cycle as a framework and the questions for each of the steps as support. In order to complement their analyses with didactical knowledge, they received an extract of the description of this error given by Padberg and Benz (2011). Each pair met another pair and completed their analyses with this new information. Finally, a plenary discussion about their analyses and about the helpfulness of the error analysis cycle was held. In order to deepen the idea of the relevance of understanding student's thinking to promote learning, preservice teachers obtained a text to read as homework about the ways in which children think differently (Selter & Spiegel, 1997).

In the second session, the focus was on the first two stages of the error analysis cycle, i.e., identifying students' errors and hypothesizing about causes for the errors. Participants were asked to identify errors in samples of students' work, interpret students thinking and make hypotheses about possible causes for the errors.

First, they commented on the main ideas and relevant examples from the homework text on how children think differently. A central point in the discussion was the need for teachers to be flexible in their own thinking in order to be able to

understand students' different ways to reason about mathematics. Preservice teachers worked in pairs analyzing samples of students' written work on fractions
with the aid of a worksheet that contained both the cases to be analyzed and
questions that guided the interpretation of the errors, asked preservice teachers to
solve similar tasks using student's erroneous reasoning and aided the search for
causes. Three cases were provided. The first was an error on the representation
of fractions, in which the areas drawn were not equivalent. This error is shown
in Figure 3.1 and has been documented and discussed by several authors (see, for
example, Ashlock, 2010; Baroody & Hume, 1991; Watanabe, 2002).

4. Use the figures to show each fraction. You will need to subdivide and shade each figure.

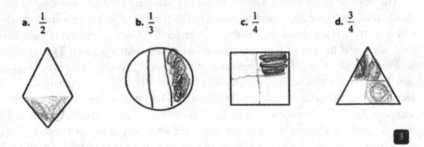

Figure 3.1   Error on the representation of fractions

The second error included in the worksheet was one of the most often reported
errors about the addition of fractions, namely adding both the numerators and the
denominators, $\frac{a}{b} + \frac{c}{d} = \frac{a+c}{b+d}$. This error, shown in Figure 3.2, has been described
and discussed by Ashlock (2010), Carpenter, Coburn, Reys and Wilson (1976),
Padberg (2002) and Rico (1995), among others.

The third case presented an error on the addition of fractions with different
denominators in which the student had added the numerators and multiplied the
denominators, as shown in Figure 3.3. This error is explained by Ashlock (2010)
as a similar error to that in the second case but with a different reasoning from
the student.

After commenting on the analyses in plenary, participants watched a video
showing a classroom situation that complemented the second case above by showing the interaction of the student with the teacher. In pairs, they commented
on the video focusing on completing the interpretation and understanding of the

1. Show with figures. Then, add the fractions.

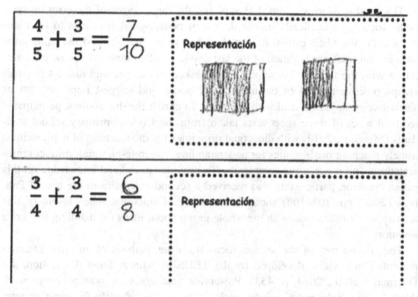

$$\frac{4}{5} + \frac{3}{5} = \frac{7}{10}$$

Representación

$$\frac{3}{4} + \frac{3}{4} = \frac{6}{8}$$

Representación

Figure 3.2 Error on the addition of fractions

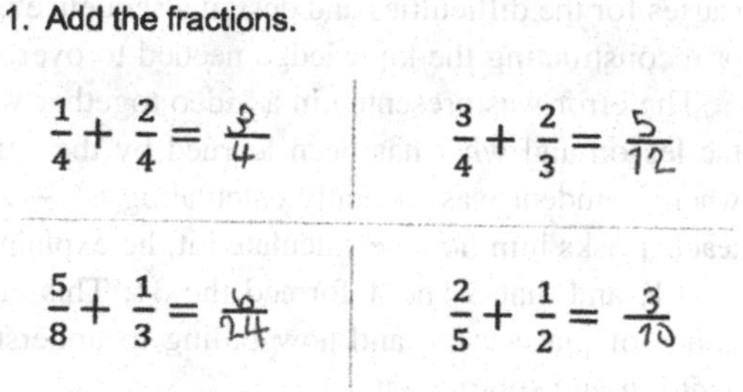

Figure 3.3 Error on the addition of fractions by multiplying denominators

1. Add the fractions.

$$\frac{1}{4} + \frac{2}{4} = \frac{3}{4} \qquad \frac{3}{4} + \frac{2}{3} = \frac{5}{12}$$

$$\frac{5}{8} + \frac{1}{3} = \frac{6}{24} \qquad \frac{2}{5} + \frac{1}{2} = \frac{3}{10}$$

misconception they had made before. Then, they were asked to form groups of four and interchange ideas. To complement this final analysis with knowledge from the literature, groups received the description of this error from Padberg (2002, pp. 102–104) and were encouraged to enrich their analyses. To finalize, interpretations and possible causes for the errors were discussed in plenary and a brainstorming of ideas on how to deal with the error was conducted as a preparation for the next session.

The third session was aimed at applying the three stages of the error analysis cycle, focusing specifically on the design of pedagogical strategies to help students overcome their errors. It started with a video that showed the same error situation on fractions addition of the last session and continued to show the interaction with the teacher. Participants were asked to analyze and discuss in small groups positive aspects of the teacher's response and suggest improvements or alternative strategies to deal with the error. To enrich the discussions, participants received a set of three short texts taken from the Chilean primary school standards (Mineduc, 2012) with didactical principles for the teaching of mathematics, namely teaching mathematics for understanding, concrete-pictorial-abstract representations and progressive complexity. To provide specialized knowledge on this particular error, participants also received a second extract from the text of Padberg (2002, pp. 105–106) about the teaching of fractions. The closure of this activity was a discussion with the whole group about ideas for handling this error situation.

The second part of the session focused on the analysis of an error situation presented in a video developed by the TEDS-FU study (for a description, see Döhrmann et al., 2014, p. 453). Preservice teachers were now asked to work through the whole error analysis cycle. They had to identify the error pattern, the underlying concepts and procedures not understood by the student and the areas of strength and difficulty. Then they had to make hypotheses about possible causes for the difficulties and design a sequence of activities aimed at constructing or reconstructing the knowledge needed to overcome the error.

The error was presented in a video together with contextual information about the lesson and what has been learned by the student's class. The error occurred when a student was mentally calculating $81 - 25$ and he answers 31. When the teacher asks him how he calculated it, he explains he thought $8 - 5 = 3$ and $2 - 1 = 1$, and that 3 and 1 formed the 31. This error situation allowed discussing issues of place value and how failing to understand them may lead to errors in addition and subtraction.

After the error was analyzed in pairs, small groups were formed to interchange their analyses and complement their views on how to interpret the error and what strategies may be appropriate to handle it. Finally, the most relevant aspects of the analysis were discussed with the whole group.

The fourth session gave preservice teachers the opportunity to use the error analysis cycle as a framework to analyze student's work, to communicate their analyses and evaluate those of others. Each group of four preservice teachers

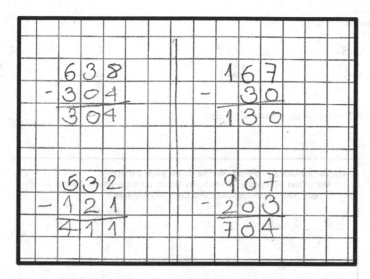

Figure 3.4 Error in subtractions with zeros

received a sample of written work from a student showing an error that they should analyze using the cycle. They were asked to do it systematically and register their ideas on a poster to be presented to the other groups later.

The six errors selected for this session were all related to operations with whole numbers and were taken from the literature on primary mathematics education. Three of them were errors in subtraction with different explanatory causes. The error in Figure 3.4 is related to subtraction including zeros, in which the student solves $n - 0 = 0$, and has been documented by Ashlock (2010), Gerster (1982), Götze et al. (2019) and Padberg and Benz (2011). The error in Figure 3.5 has been described by Ashlock (2010) and Rico (1995) and involves problems with the regrouping process in subtraction. Figure 3.6 shows an error on horizontal subtraction that is associated with a lack of understanding of place value and has been found in a study by Götze et al. (2019) and by Lucchini, Cuadrado and Tapia (2006). The latter authors suggest that students who subtract this way, without considering place value, often do their calculations based just on the digits.

The other three tasks involved errors in multiplication. The error in Figure 3.7 has been documented by Ashlock (2010) and reveals confusion in the multiplication algorithm when there are two-digit multipliers. The second multiplication error (Figure 3.8) occurs when a student applies the algorithm multiplying from

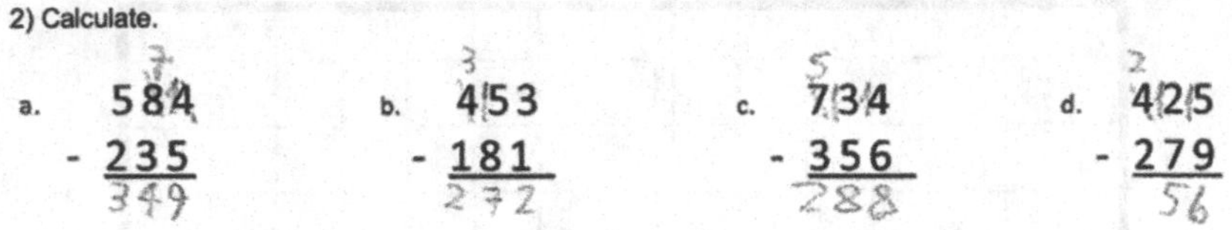

2) Calculate.

a. $5\,8\overset{7}{4}$ b. $4\overset{3}{|}5\,3$ c. $7\overset{5}{|}3\,4$ d. $4\overset{2}{|}2\,|5$
     $-\ 2\,3\,5$ $-\ 1\,8\,1$ $-\ 3\,5\,6$ $-\ 2\,7\,9$
     349 $2\,7\,2$ 288 56

Figure 3.5 Error in subtraction algorithm

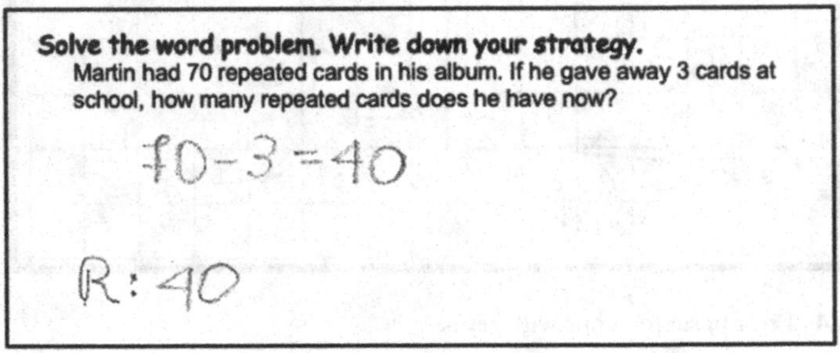

Solve the word problem. Write down your strategy.
Martin had 70 repeated cards in his album. If he gave away 3 cards at school, how many repeated cards does he have now?

$70-3=40$

$R: 40$

Figure 3.6 Place value error in subtraction

right to left but writes down the result from left to right, not accounting for place value. This error has been reported by Lucchini et al. (2006). The last error, displayed in Figure 3.9, has been informed by Padberg and Benz (2010) and by Götze et al. (2019). It consists in missing to register some of the partial products when using the distributive property to solve a multiplication in which both factors have two digits.

After each group discussed the error they received and registered their analysis in the poster, the presentation phase took place. Two members of each group stayed at their poster to explain their analysis to other participants and the two remaining members went around observing and evaluating the other groups' analyses and proposals. After a while, group members changed roles. This allowed all participants to communicate and explain their own error analysis and to reflect on those made by other groups, expanding their repertoire of errors and widening their views on possibilities for analysis.

To finalize the session and the seminar sequence, there was time for whole group discussion. There was time both for discussion of the error analyses of the poster presentations and to reflect on what was learned during the seminar sequence and to formulate questions and concerns that remained at the time still open.

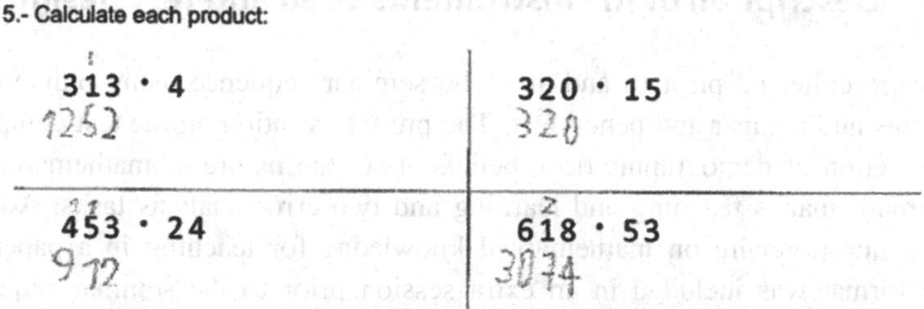

5.- Calculate each product:

$$313 \cdot 4$$
$$1252$$

$$320 \cdot 15$$
$$328$$

$$453 \cdot 24$$
$$912$$

$$618 \cdot 53$$
$$3074$$

Figure 3.7 Error in two-digit-factor multiplication algorithm

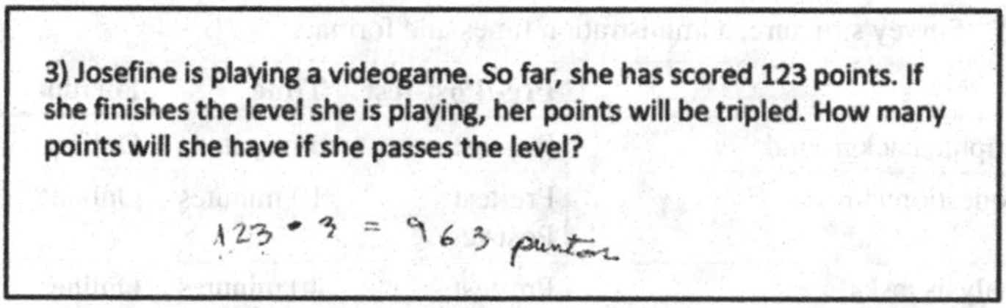

3) Josefine is playing a videogame. So far, she has scored 123 points. If she finishes the level she is playing, her points will be tripled. How many points will she have if she passes the level?

$$123 \cdot 3 = 963 \; puntos$$

Figure 3.8 Place value error in multiplication

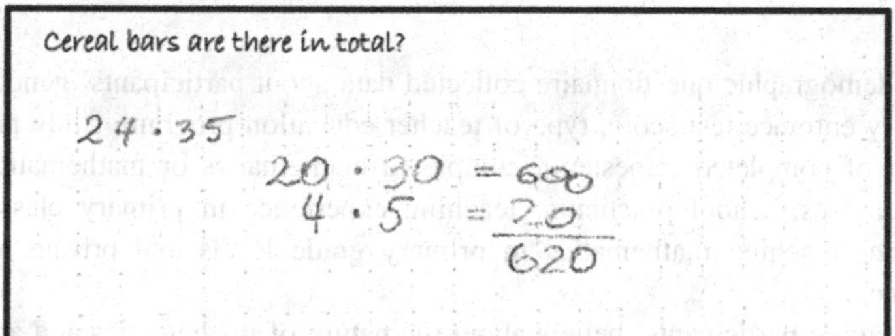

Cereal bars are there in total?

$$24 \cdot 35$$
$$20 \cdot 30 = 600$$
$$4 \cdot 5 = \underline{20}$$
$$620$$

Figure 3.9 Error by decomposing multiplication

3.1.2   Description of the Instruments Used and Test Design

Data were collected prior to and after the seminar sequence using online questionnaires and a paper and pencil test. The pre-intervention online test comprised
the collection of demographic data, beliefs about the nature of mathematics and
about mathematics teaching and learning and two error analysis tasks. Additionally, a questionnaire on mathematical knowledge for teaching in a paper and
pencil format was included in an extra session prior to the seminar sequence.
The post-intervention test also contained two error analysis tasks, the beliefs
questionnaires and a short survey about participation in the seminar sequence.
Table 3.1 shows the structure of the tests and the time allocated for them.

Table 3.1 Survey structure, administration times and formats

Part	Pre-/Post-test	Time	Format
Demographic background	Pre-test	5 minutes	Online
Beliefs questionnaires	Pre-test Post-test	10 minutes	Online
Error analysis tasks	Pre-test Post-test	40 minutes	Online
Final information	Post-test	2 minutes	Online
Mathematical Knowledge for Teaching	Pre-test	60 minutes	Paper and pencil

The demographic questionnaire collected data about participants' gender, age,
university entrance test score, type of teacher education program, study progress
(number of completed semesters), completed mathematics or mathematics education courses, school practicum, teaching experience in primary classrooms,
experience teaching mathematics in primary grade levels and private tutoring
experience.

Data about participants' beliefs about the nature of mathematics and about its
teaching and learning were also collected. They had to express their agreement or
disagreement on a six-point Likert scale to a series of statements related to these
two constructs. The items were taken from those used in the beliefs questionnaires
of the TEDS-M study (Tatto et al., 2008), which considered two scales for the
assessment of each of the beliefs areas named above. In particular, the Chilean
version of the items was used (Ávalos & Matus, 2010). The area of beliefs about
the nature of mathematics contemplated a scale viewing this discipline as a set of
rules and procedures that have to be learned and strictly applied to find the correct

answers to mathematical problems and another scale viewing mathematics as a discipline focused on inquiry processes, in which mathematical procedures are tools for solving problems that can be flexibly applied. The questionnaire on the beliefs about learning mathematics was formed by two scales too. On the one side, there were items related to a teacher-centered view of learning mathematics, in which teachers pass on information, formulas and procedures to students, who have to memorize and apply them to solve mathematical tasks. On the other side, a scale related to a student-centered view of learning mathematics was used; it implies that students must be active in order to learn mathematics, they should be involved in inquiries and development of own strategies to solve mathematical problems and they must understand the concepts and procedures they apply.

To assess preservice teachers' specialized knowledge, a multiple-choice Mathematical Knowledge for Teaching test in a paper and pencil format was applied. The items of this test came from the Learning Mathematics for Teaching Project (Hill et al., 2008) and were adapted and validated to be used with preservice Chilean teachers in the Refip project (Martínez et al., 2014). In particular, the test booklet aimed at assessing specialized knowledge in the area of numbers and operations for preservice primary school teachers was used because the errors chosen for the sessions and for the pre- and post-test were from this domain of mathematics as well. The instrument had 24 questions, from which 15 were multiple-choice and 9 were complex multiple-choice queries comprising 3 or 4 items each. In total, the survey had 45 items, that respondents had to complete in 60 minutes.

For the error analysis tasks, four error situations were chosen. For all of them, items were developed following the same structure. First, some contextual information was provided, including the grade level, the content of the lesson and what has been done in the class where the error occurs. Next, the error situation is presented in a short video clip. The video can be watched many times until the respondent goes to the next page, where a picture of the written error is displayed. Respondents are asked to briefly describe the student's error and to answer similar tasks using the student's erroneous thinking. Figure 3.10 exhibits this part of the test for a whole-numbers addition error.

On the next page, the erroneous procedure used by the student was revealed so all respondents could continue answering the questions, including those who had not identified the error in the first instance. The online questionnaire was programmed so that after answering questions on a page and clicking on "next", it was not possible to go back and change previous answers or watch the video again. Answers were saved to the database every time "next" was clicked on. This

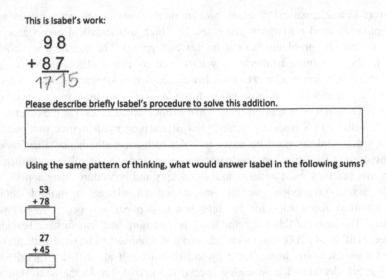

This is Isabel's work:

$$98$$
$$+\ 87$$
$$\overline{17\,15}$$

Please describe briefly Isabel's procedure to solve this addition.

Using the same pattern of thinking, what would answer Isabel in the following sums?

53
+ 78
☐

27
+ 45
☐

Figure 3.10   Sample item for the identification of student error

allowed to provide information that revealed answers from questions on previous pages but that was important to continue with the survey.

The question on the next page of the survey suggested a series of causes for the error. For each of them, respondents had to decide if it was a possible cause for that error situation, as shown in Figure 3.11. Here, it was again relevant that the respondents could not go back to the previous page of the test.

After participants decided on the plausibility of a list of hypotheses about the causes for the error situation, a brief explanation of the causes considered as correct for was given on a different page. Respondents were asked to, based on this given diagnosis, propose three alternatives to help the student recognize their error and overcome it. Figure 3.12 shows a sample of this page of the test.

Lastly, the final information questionnaire asked respondents about the sessions they attended and their degree of involvement in the seminar sequence. In this item, they had to select if they were "very active", "active", "little active" or "barely active" during the sessions.

Data collection took part in two formats. All questionnaires were online-based, with exception of the Mathematical Knowledge for Teaching test, which was applied in a paper and pencil format. Whilst the demographic background and beliefs questionnaires and the Mathematical Knowledge for Teaching test were applied

For each of the following statements decide if it is a possible cause for Isabel's error or it is not.

	It is a possible cause	It is not a possible cause
Isabel didn't read the exercise properly.		
Isabel has an inadequate idea of place value.		
Isabel doesn't know well the basic addition facts		
Isabel has misunderstood the addition algorithm.		
Isabel doesn't know how to use regrouping in the addition algorithm.		
Isabel takes each column as a separate addition.		
Isabel has forgotten de commutative property of addition.		
Isabel applies the addition algorithm from left to right. She starts adding from the tens.		
Isabel has not understood the relationship between the number one that should be written on the tens column and regrouping and place value.		
Isabel forgot to write down the number one in the tens column.		

Figure 3.11   Sample close-ended item for the hypothesizing about causes of student error

in the same way for all participants, the error analysis test followed a particular design.

The two error analysis tasks included in each testing time were selected from a pool of four tasks. This allowed developing four different test versions with different combinations of tasks (see Table 3.2) so that the difficulties of the tasks could be related across the test booklets.

To ensure that respondents only answered each task once and thus avoid gains in the post-intervention tasks by familiarity with the cases, an application scheme was designed. In the pre-test, respondents were randomly assigned to one of the four test booklets. Depending on the allocation in the pre-test, a booklet version

You discover Isabel has difficulties to understand the regrouping process and how it is recorded in the traditional addition algorithm.
Considering this diagnosis, you want to help Isabel to recognize the error on the procedure she uses.
Please suggest three different possibilities to do this.

Possibility 1

Possibility 2

Possibility 3

Figure 3.12 Sample open-ended item for error handling

Table 3.2 Error analysis tasks distribution in the survey booklets

Booklet	Task 1	Task 2
Booklet 1	Subtraction with inversion task	Addition algorithm task
Booklet 2	Subtraction with inversion task	Numbers transcoding task
Booklet 3	Subtraction with zero task	Numbers transcoding task
Booklet 4	Subtraction with zero task	Addition algorithm task

was allocated in the post-test, which contained exactly the two tasks not yet processed. For instance, a preservice teacher answering Booklet 1 before the seminar sequence would receive Booklet 3 at the post-test.

3.1.3   Description of the Tasks

The four error situations selected for the error-analysis-tasks were from primary school mathematics in the area of number and operations. One of them showed an error when transcoding numbers, two were errors in the subtraction algorithm and one in the addition algorithm. The following sections describe each of the errors from a mathematics education perspective.

3.1.3.1 Number Transcoding Error

Successful mastery of the place-value structure is a major and fundamental goal of primary school mathematics. In fact, it has been shown that understanding the composition of our numerical system is key for developing more complex arithmetic competencies and that early deficits in this knowledge are related to later difficulties with multi-digit calculations (Moeller, Pixner, Zuber, Kaufmann & Nuerk, 2011).

Children's difficulties in acquiring place-value knowledge have been reported by several studies. One particular difficulty is related to learning the correspondence between numbers given verbally and written Arabic numerals (e.g., Cayton & Brizuela, 2008; Granà, Lochy, Girelli, Seron & Semenza, 2003; Power & Dal Martello, 1990). In the Spanish language, as in most Western languages, the verbal number system has particular characteristics that have to be learned by children in order to master this oral-written number correspondence. Power and Dal Martello (1990) suggested that an internal abstract semantic representation is constructed when verbal numbers are translated into Arabic numerals. This representation is closely related to the verbal codification and is composed of primitive numerical concepts and multiplicative and additive syntactic rules. Primitive concepts (noted with a C) include units (0 to 9) and powers of ten (10, 100, 1000, up to 1 million). Every non-primitive number can be built by applying additive and multiplicative rules to combinations of the primitive number. For instance, the number "*seventy-nine*" would be represented by the semantic expression ((C10 * C7) + C9) and the number "*two hundred seventy-nine*" would be represented by the semantic expression ((C100 * C2) + (C10 * C7) + C9). The authors suggest this translation is done in two stages. It starts with the interpretation of the verbal number and the construction of its semantic representation, which is then used to produce the Arabic numeral. The application of two rules is also implied in this process, i.e., by addition relations overwriting of zeros has to be done and by multiplicative relations concatenation has to take place (Granà et al., 2003).

Number dictation has been used to examine students' number knowledge skills and identify patterns of errors that may provide hints of their (mis)understandings. Power and Dal Martello (1990) found that most 6–8 years-old children were able to correctly interpret and register two-digit numerals, but they struggled with three- and four-digit numbers. They identified a dominance of syntactic errors in which additional zeros were inserted (e.g., 279 was written as 20079 or 2079). They suggested that this occurred when children have not yet mastered the overwriting rule and they use the concatenation rule instead. For example, transcoding the number word "*two hundred and nine*" requires overwriting a number 9 on the zero at the units' place in 200. Instead of doing this, children use the concatenation rule and write the nine at the right of the 200, which yields 2009.

From an intercultural study, Scheuer, Sinclair, Merlo de Rivas and Tièche Christinat (2000) concluded that syntactic errors are part of a developmental process of numerical notation skills, in which conventional and non-conventional notations co-exist. During this process, children first master conventional notation of two-digit numbers. The understanding of verbal quantities greater than one hundred precedes its conventional notation. This can be seen in that children are usually able to represent verbal numbers in an additive system (with coins or other base-ten material) before they can write the Arabic number correctly. The challenge imposed by the place-value notational system acquires here evident relevance.

Scheuer and colleagues (2000) also point out that when students used non-conventional notations, showing syntactic errors, they focused mainly on registering every number-word included in the verbal expression and tended to forget other characteristics of numbers such as numbers 100–999 are written with three digits. Lerner, Sadovsky and Wolman (1997) add to this point by stating that students who can write conventional two-digit numbers usually show syntactic errors in numbers with hundreds, and children who can already write correctly two- and three-digit numbers exhibit syntactic errors in numbers from a thousand.

Orozco-Hormaza, Guerrero-López and Otálora (2007), in a study with Spanish-speaking children, developed a classification of syntactic errors. They suggest that during the interpretation phase, verbal numbers are decomposed into fragments by applying previously learned rules. In the production phase, one numeral for each fragment is considered and they are put together using three types of relations: juxtaposition, compaction or concatenation. In the juxtaposition case, numbers coding each of the fragments are written next to each other. For example, the number 279 (*doscientos setenta y nueve*) could be written as 200709 if the student is fragmenting it in three parts or as 20079 if *seventy-nine* is already taken as an element on its own. Alternatively, when compaction is used, the inferior-order fragment is overwriting the zero in the immediately superior-order fragment, e.g., 279 (*doscientos setenta y nueve*) is written as 2079. Lastly, when a student uses the concatenation strategy, she registers only the digits indicating the quantity indicators in the verbal numeral. In this case, 209 (*doscientos nueve*) would be written as 29, excluding the zeros required to maintain place value.

The "Number transcoding error task" included in the survey consisted of a student in a number-dictation situation. He shows a syntactic error when attempting to write the Arabic numerals. The errors are displayed in Figure 3.13, which shows the dictated numbers on the left column and the numbers written by the student on the right column.

Figure 3.13 Number transcoding error

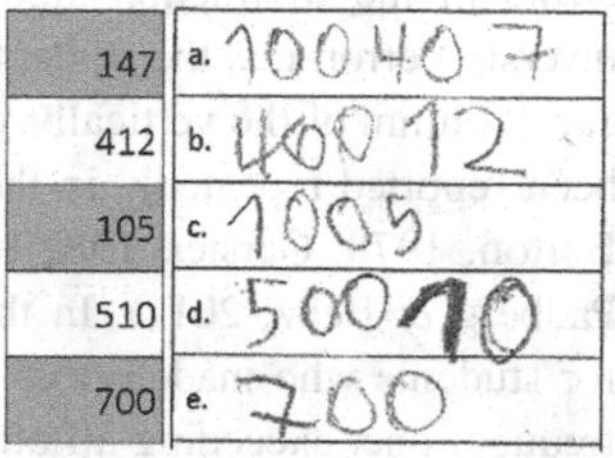

147	a. 100407
412	b. 40012
105	c. 1005
510	d. 50010
700	e. 700

3.1.3.2 Subtraction Algorithm Errors

Students' errors in written subtraction are very common among primary-school children, even more than those related to addition. Cox found that "over twice the number of students made systematic errors in subtraction than in addition" (1974, p. 48) and that the percentage of these difficulties were greater in tasks requiring regrouping.

At least two subtraction algorithms have been recognized as widely taught in schools: the *taking away* method (also known as decomposition method) and the *determining the difference* method (also known as the Austrian subtraction method). The former method is to subtract the number on the minuend to the number on the subtrahend. When a digit in a place-value column is smaller than the digit in the subtrahend, one group of the next-bigger place-value column is regrouped in ten smaller units; these units, together with the original ones, are then sufficient to take away from and the next-bigger place-value column is reduced by one. The latter model functions with a different strategy, which starts from the subtrahend and the difference to get to the minuend is looked for. For further details on both methods, see for example, Selter, Prediger, Nührenbörger and Hußmann (2012) or Jensen and Gasteiger (2019). In the present study, errors more closely associated with the *taking away* method were considered because this is the method most widely used in Chilean schools.

Several studies have reported that subtractions requiring regrouping are the tasks causing the most difficulties for students using the *taking away* method (Fiori & Zuccheri, 2005; Jensen & Gasteiger, 2019). Because in this method the relationships between columns are needed to carry out the procedure, the understanding of the place value structure is of particular significance. This is of special relevance in tasks in which digits in the minuend are smaller than the ones in the subtrahend and regrouping is necessary (Fuson, 1990; Roa, 2001; Sánchez & López Fernández, 2011).

Two of the four error analysis tasks included in the survey in this study showed errors in the subtraction algorithm requiring regrouping. One of them was an inversion error, i.e., the student subtracted the smaller digit from the larger within each column of the vertically written subtraction (see Figure 3.14). This error has been reported repeatedly in the field (see for example, Ashlock, 2010; Brown & Burton, 1978; Gerster, 1982; Götze et al., 2019; Kühnhold & Padberg, 1986 or Padberg & Benz, 2011). In their study, Cebulski and Bucher (1986) found that for students who made errors on every problem, this type of error was the most frequent one, exceeding all other error types combined.

In the search for causes, a vertical (mental) inversion of the digits in order to be able to subtract the smaller from the larger has been interpreted as a repair strategy when the student is not capable of executing the regrouping procedure (Brown & Van Lehn, 1980). Ashlock (2010) suggested that students showing this error pattern may be considering each place-value position as separate subtraction problems, that they may not be thinking of minuend and subtrahend as numbers on their own. Moreover, he implied that students may be overgeneralizing the commutative property of addition to subtraction. He also stated that this error may be rooted in statements heard by the student, such as 'always take away the smaller from the larger' or 'it is important to maintain the columns when subtracting'.

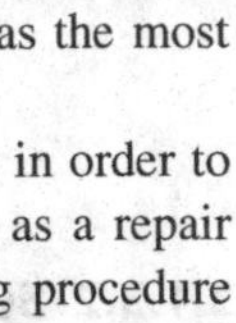

Figure 3.14 Subtraction with inversion error

$$
\begin{array}{r}
524 \\
-298 \\
\hline
374
\end{array}
$$

The second subtraction error analysis task occurs when there is a zero digit in the minuend and regrouping is required. In other words, in a column of the algorithm $0 - N$ is found and the student, instead of doing the borrowing or regrouping, writes N as the answer for that column (see Figure 3.15). This error has also been repeatedly reported in the literature (Brown & Van Lehn, 1980; Gerster, 1982; Götze et al., 2019; Roa, 2001). Young and O'Shea (1981) argue that this error may be part of the inversion error above and, therefore, may have similar causes. However, they add, it is also sometimes a pattern of error by itself because some children who are capable of regrouping and who do not show the inversion error nevertheless show this type of error when there are zeros in the minuend. Additionally, difficulties in understanding the meaning of zero, relating it to 'nothing' and failing to interpret its relevance as a place-value holder have also been suggested as causes underlying this error (Fiori & Zuccheri, 2005; Sánchez & López Fernández, 2011).

Figure 3.15 Subtraction with zero error

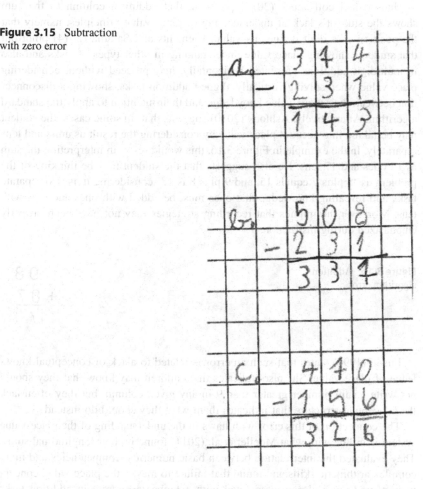

3.1.3.3 Addition Algorithm Error

Errors in the written addition algorithm can take many different forms, accounting for various erroneous strategies, including not carrying out a necessary regrouping or doing it wrongly. One of the most common errors occurs when, as shown in Figure 3.16, students write the regrouping digit as an additional column in the result (Cox, 1974; Gerster, 1982; Hansen et al., 2017; Padberg & Benz, 2011). This was the fourth error included for analysis in the survey of the present study.

Hansen and colleagues (2017) postulate that adding a column to the sum shows the student's lack of understanding of place value principles, namely that the position of a digit defines its value. Clements and Sarama (2014) point out that students showing place value understanding in other types of tasks and also in multidigit additions presented horizontally, may proceed without considering place value when solving vertically aligned addition tasks, showing a disconnection between their place-value knowledge and their intention to apply the standard algorithm. Alternatively, Ashlock (2010) suggests that, in some cases, the student may be taking into account place value by considering the result as units and tens separately. In the example in Figure 3.16, this would result in interpreting the sum as 15 ones and 17 tens. He also suggests that the student may be thinking of the problem as '8 plus 7 equals 15, and 9 plus 8 is 17' considering it as two separate tasks and revealing knowledge that ones must be added with ones and tens with tens. Moreover, he implies that recording strategies may not have been properly emphasized during instruction.

Figure 3.16 Addition
algorithm error

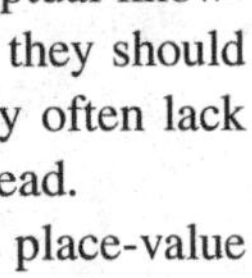

$$
\begin{array}{r}
9\,8 \\
+\ 8\,7 \\
\hline
1715
\end{array}
$$

Fuson (1990) states that such an error is related to alack of conceptual knowledge of place value. She also indicates that children may know that they should not write a sum number greater than 9 in any given column, but they often lack the conceptual structure that indicates them what they should do instead.

The connection of this error with flaws in the understanding of the place-value system is in line with what Moeller et al. (2011) found in their longitudinal study. They evaluated the interrelation between basic numerical competencies and more complex arithmetic skills and found that failure to master the place-value concept in first grade was closely associated with a higher error rate in addition tasks requiring regrouping.

3.2   Data Collection

Preservice teachers from 11 different universities from Santiago, Chile, participated in the university seminar sequence. In three universities the seminar sequence on error analysis was embedded into their regular academic activities. Participation in the survey was voluntary. In a fourth university, an open extracurricular seminar sequence was offered, in which preservice teachers from 8 different universities took part. A total of 165 preservice primary teachers and 30 preservice special teachers participated in the seminar sequence, from which 131 answered both pre- and post-test surveys.

In Chile, primary teachers' initial education usually has a duration of 8 to 10 semesters. The formation focus is on school grades 1 to 4 or 1 to 6. In some universities, students are offered the opportunity to widen their knowledge and extend their working possibilities up to grade 8 in some particular subject. Students taking this option usually have additional courses equivalent in duration to two semesters, in which they focus on contents and pedagogical issues of higher (low-secondary) grades. This is called a specialization. Because participation in such learning opportunities is expected to have an impact on preservice teachers' knowledge about mathematics and mathematics education, the distribution of participants in this kind of program will be considered.

Also of relevance is the advancement of studies of the participants. A higher number of completed semesters at the teacher education program is associated with more opportunities to learn about pedagogy, mathematics and mathematics education and to develop associated competences. Moreover, Chilean teacher education programs usually include school-based placements already at the beginning of the program, starting with a very passive, observing role and moving to a more active one with the advancement within the program. This results in that students of higher semesters usually have had more practical experiences and thus more chances to reflect on teaching and learning of mathematics and to develop associated competencies.

Table 3.3 shows the frequencies of participants following three different courses of studies: special education, primary teacher education without specialization in mathematics (including those with specializations in other subjects) and primary teacher education with specialization in mathematics for each year of advancement in the course of studies.

Table 3.3 Participants' year and course of studies

Course of studies	Year 1	Year 2	Year 3	Year 4	Year 5	Total
Special Education	16	0	0	0	1	17
Primary teacher education	1	0	56	13	3	73
Primary teacher education with specialization in mathematics	15	0	6	4	16	41
Total	32	0	62	17	20	131

3.3   Data Analysis

The survey included both closed and open-ended items. The different types of answers given by participants were analyzed using qualitative and quantitative methods. In the following subsections, these methods will be described.

3.3.1 Qualitative Data Analysis

Among other types of items, open-ended questions were used in the survey of this study to assess preservice teachers' diagnostic competence in error situations. In particular, open items were used to investigate the causes preservice teachers can hypothesize for the error situations and how they would pedagogically respond. The answers given by participants to such items are short pieces of text. In one of them, they suggest three possible causes for the student error presented in the task and in the other, they describe how they would proceed to help the student overcome the error. Although brief, the answers showed a wide variety of ways of communication, with different levels of detail and deepness and diverse writing styles. This richness and complexity are fundamental characteristics of qualitative data, which also require complex and systematic analysis techniques.

Various methods and approaches for interpreting and analyzing text data have been developed. For this study, qualitative text analysis was chosen because it provides a framework to interpret and reduce the complexity of the answers so their interesting and relevant characteristics can be considered for analysis in relation to the research question. To achieve this, qualitative text analysis works with categories. They are, according to Kuckartz (2019, p. 183), "of crucial importance for effective research, not only in their role as analysis tools, but also insofar as they form the substance of the research and the building blocks of the theory the researchers want to develop." Kuckartz (2014) distinguishes between five types of categories in empirical social science research:

- *Factual categories*: designate objective or assumingly objective features, such as occupation, area of expertise or teaching subject.
- *Thematic categories*: denote a specific theme, argument or matter. Different passages referring to particular contents are then categorized accordingly.
- *Evaluative categories*: a prescribed number of characteristics, usually ordinal levels, are used to rate the information. For instance, the category 'student participation' could be classified as 'very active', 'active', or 'passive'.
- *Formal categories*: refer to information about the units being analyzed, such as the date of the interview or its length.
- *Analytical or theoretical categories*: they are part of subsequent work with descriptive categories. Categories that may be used to characterize the text in an initial stage may then be put together into a broader category, which would constitute an analytical category or into a category that refers to an existing theory, which would be a theoretical category.

In the present study, thematic and evaluative categories were used to code the hypotheses about causes of student errors suggested by preservice teachers in the open-ended items. Due to the complexity of the category system and the fact that their competence to hypothesize about causes was also measured using quantitative data from another item, the analysis of these coded data will be presented as a side-study in an upcoming publication.

To analyze the answers given by preservice teachers on how they would deal with the student error (see section 3.1.2 and figure 3.12 for a sample question), evaluative categories were defined. This results from the intention to focus on different features of the decisions preservice teachers would make in facing the errors and the degrees in which these aspects were manifested. The questions on the survey asked preservice teachers to provide three different alternatives to deal with the student's error. Each of these suggestions was considered a text segment to which a category should be assigned. The characteristics of evaluative qualitative text analysis and how it was applied in this study will be presented in the next section.

Evaluative Qualitative Text Analysis
Evaluative qualitative text analysis "involves *assessing, classifying, and evaluating content*" (Kuckartz, 2014, p. 88). Researchers assess the data and build

categories. Characteristics for these categories are then defined and usually recorded as ordinal levels, which may be later used to explore associations and test hypotheses.

To ensure its quality, the method is a systematic and rule-governed process, for which Kuckartz (2014) suggests seven phases. Each phase is meant for a single evaluative category. Because multiple categories were evaluated in the present study, phases 2 to 5 were repeated for each category. The seven phases always have the research question at its core and are described by Kuckartz (2014) as:

1. Define evaluative categories
2. Identify and code the text passages that are relevant for the evaluative category in question
3. Compile the text segments coded with the same code
4. Define levels (values) for the evaluative categories and assign them to the text segments. If necessary, modify the category definition and the number of category values
5. Evaluate and code the entire data set
6. Analyse and present results 1: Category-based analysis
7. Analyse and present results 2: Overviews (qualitative and quantitative), in-depth interpretation of cases. (Kuckartz, 2014, p. 89)

Qualitative text analysis methods are not to be automatically applied. According to Mayring (2015), they have to be adapted to the particular characteristics of each study, so they are appropriate to the material to be evaluated and to the particular context and research question involved. The particular form of using qualitative text analysis to evaluate data in this study will be detailed below.

The process of evaluative text analysis started with the definition of four categories, coming from the research question and the theoretical framework and focusing on two aspects of the suggestions given by preservice teachers to deal with students' errors. Specifically, it was of interest to evaluate if their pedagogical proposals considered the student as an active learner by focusing on what the student would do, think or say or, on the other hand, were based on a teacher-directed approach, in which the learning process is focused on what the teacher does to pass on information to students, who have to memorize and apply. This would allow looking for associations between the decisions preservice teachers make to deal with student errors and their beliefs about teaching and learning of mathematics. A second relevant aspect to assess was the type of pedagogical goal of the suggestions to dealing with student errors. There have been animated discussions in the past decades in the field of primary mathematics education about the need to move from an exclusive focus on mastering mathematical procedures towards teaching mathematics for understanding (see discussion about conceptual and procedural understanding in section 1.3.3.3). Thus, it was interesting to investigate to what extent preservice teachers' decisions to dealing with student errors would aim to promote conceptual understanding, procedural understanding or a mixture of both.

An initial examination of the responses confirmed that both pairs of categories were relevant and to be found in preservice teachers' suggestions for dealing with the errors. It also confirmed the need to evaluate their occurrence using ordinal levels. Additionally, a fifth category arose from this first review, namely the need to distinguish to what extent the strategies suggested by preservice teachers gave evidence of being targeted to address the particular error, in the sense of suggesting strategies that would account for what the particular student already knows and take this as a starting point to promote further learning.

An additional aspect that needed to be considered is the segmentation of the text to be coded. To approach the material, the structure of the survey was taken into account, so each of the three pedagogical suggestions to overcome the student's error (each answer) was considered a single unit to be coded. To account for the fact that some respondents made some connections between the three answers, they were kept together and coded sequentially so they may act as context units and included as reference in the analysis when implicit indications to previous answers were made. Thus, each answer was taken as a unit of analysis and the other two were also available to be checked for a complete interpretation when needed. Regarding the sequence, the interpretation of the answers was made following their position in the questionnaire.

After the relevant text passages had been identified and compiled and before proceeding to the data evaluation phase, coding manuals were developed. According to Mayring (2015), defining a category precisely requires consideration of three aspects, including a description of the text components belonging to the category, anchor examples extracted from the data that illustrate what should belong to the category and coding rules that refer specifically to border cases in which the limits of a particular category may be problematic. In evaluative qualitative text analysis, the values or levels for the categories are critical. Deciding how detailed the category distinctions can be, depends on the research question and on the characteristics of the text passages to be evaluated. For each of the five categories in this study, it was differentiated between four levels: strong evidence of the category, some evidence of the presence of the category, no evidence of the category and unable to classify, usually meaning that the answer is incomplete, unclear, ambiguous or contradictory and, therefore, cannot be classified.

To test the coding manual, 20% of the answers were coded. Critical points requiring more precise definitions and rules to address problems of delineation of values were identified and accounted for in a second version of the coding manual. At this stage, it was of particular relevance to provide examples for each of the error tasks as, in some points, there were nuances in the interpretation of the categories for each of the error tasks.

The next stage consisted of evaluating the data. To ensure the quality of the coding process, the data were evaluated by two coders. In addition to the main researcher, a second coder was trained prior to the beginning of the coding process, so both had a good understanding of the research question, the theoretical background and the meaning and delimitations of the categories. A consensual coding approach was taken, aimed at reducing as much as possible coding differences by discussing conflicting interpretations and finding agreement (Kuckartz, 2014). Using the developed coding manual, 20% of the answers were independently coded. Then, similarities and differences were discussed to look for sources of error. For differences, arguments and reasoning were shared until discrepancies were clarified and a consensus was reached. These discussions led to further refinement of the coding manual, the category and level definitions and their examples. At this point, the focus was to understand and interpret sources of irregularities. With this refined version of the coding manual, one-third of the data was evaluated and the process of refinement was repeated. A smaller number of differences arose this time. On a third cycle, an independent coding approach was taken, and data were evaluated separately by the two coders.

Although qualitative text analysis is systematic and follows a set of rules and despite the efforts to describe precise categories, it is a qualitative method involving the interpretation of data. Hence, an important quality criterion is intercoder agreement, which measures how strongly multiple coders agree in how they interpret the same data. A tool to increase this agreement is a detailed and worked-out coding manual that provides precise definitions and examples and leaves less room for divergent interpretations. However, interpretation is inherent to the coding process, so it is not totally avoidable, and, in most cases, a certain degree of variation will exist (Schwarz, 2015). Although it has been argued that interrater reliability coefficients are not necessary, and appropriate procedures like the consensual coding approach described above are sufficient to ensure intercoder agreement (Kuckartz, 2014), others suggest that summarizing the intercoder reliability with a measure like Cohen's Kappa (Cohen, 1960; McHugh, 2012) is useful and, as total agreement is not to be expected, a value above 0.7 should be interpreted as sufficient (Mayring, 2000).

In the present study, in addition to using the consensual coding approach during the first stages, intercoder agreement was checked after the entire data set was evaluated using Cohen's Kappa. The reliability values for each category of the four error analysis tasks are displayed in Table 3.4. All the measures are over the 0.7 value suggested by Mayring (2000) and are therefore interpreted as adequate.

Table 3.4 Cohen's Kappa values to evaluate intercoder agreement in the coding of open-ended items

Categories	Subtraction with inversion	Subtraction with zero	Addition algorithm	Number transcoding
Active-learner approach	.733	.739	.772	.736
Teacher-directed approach	.733	.739	.772	.736
Conceptual understanding	.771	.791	.784	.762
Procedural understanding	.729	.757	.729	.703
Targeted strategies	.760	.750	.803	.711

3.3.2 Quantitative Data Analysis

Besides the open-ended items, closed items were included in the survey to evaluate preservice teachers' diagnostic competence in error situations. The questionnaire on mathematical knowledge for teaching (MKT) was also composed of closed items. Therefore, both the answers to the items focusing on the competence to hypothesize about causes for students' errors and the responses on the MKT test were analyzed using Item-Response- Theory methods (hereafter IRT). The codes resulting from the evaluative text analysis of the decision-making items were further analyzed using Latent-Class-Analysis (hereafter LCA).

The selection of two different methods for analyzing answers in the diagnostic competence survey is based on the very nature of what is being evaluated. On the one hand, it is possible to distinguish between causes that are potentially possible for the occurrence of a mathematical error and others that do not or at least reach a consensus on the matter. Expert judgment was used to complement the corresponding literature and determine the correct answers to each of the questions. On the contrary, the ways in which preservice teachers decide to deal with students' errors, that is, the pedagogical proposals they present, are difficult to judge as correct or incorrect. The appropriateness of certain teaching strategies depends on a wide variety of factors, including school culture and beliefs, characteristics of the group, of the student, the teacher and the available materials, among others. This impossibility to dichotomize between right and wrong answers raised the need to use a method different from IRT because it requires that answers are first codified into correct and incorrect. Alternatively, Latent-Class-Analysis provides the opportunity to identify patterns in preservice teachers' answers and therefore build categorical classes showing different profiles of answers to the items.

In the following sections, the fundamentals of the Item-Response-Theory methods and the way in which answers to the MKT test and the items of hypothesizing about the causes of student errors were analyzed, including expert judgment, will be described first. Next, the LCA methods used to find pedagogical decision-making profiles will be explained.

3.3.2.1 Item-Response-Theory

The measurement of unobservable variables is often of interest in psycho-social contexts but a difficult task. Because such attributes are not directly visible, they can only be evaluated through observable associated indicators. Classic Test Theory (CTT) has been used in the psychological and educational fields for many decades and is based on the assumption that each person has a true score that could be obtained if there were no errors in measurement procedures. In other words, it is founded on the idea that the observed score obtained by an individual on a test is the result of adding his or her true score and an error score. Therefore, item difficulty parameters and person scores are closely related and undiscernible. For instance, if a test is administered to high-ability individuals, items will appear to be easy; if the same test is answered by a low-ability group, the items will show higher difficulty indices. Moreover, reliability will be affected by respondents' and items' characteristics. If a homogeneous group of highly talented students takes a test, reliability measures will be weaker than if the same items were applied to a heterogeneous group. Correspondingly, high reliability measures are to be expected if a test is composed of moderately discriminating items, but lower reliability will be observed if the same group answers a test with either very low or very high discriminating values (Meyer, 2014). Therefore, under CTT, inferences are made about the scores a person obtains on a test, on a particular set of items (Wu, Tam & Jen, 2016).

Alternatively, in Item Response Theory (IRT), there is an explicitly defined latent trait that can be measured by different sets of items as long as all the items refer to the same latent trait (Wu et al., 2016). In contrast to CTT, IRT methods are able to provide information that separates characteristics of the particular test and the sample. IRT methods estimate parameters for each item on a test and also provide a person parameter for the latent trait being measured that is separated from the person's answer to the particular items (Yang & Kao, 2014). As in IRT models a latent trait is conceived to be underlying the observed answers, an important assumption is unidimensionality, which implies that all items are related to the same latent ability or trait (DeMars, 2018). There are also multidimensional IRT models, but in this study, only unidimensional models will be used as one competence is measured. A second assumption is that of local independence, which means that responses to the items should not be statistically related to each other after controlling for the latent trait. If local independence does not hold, it can be an indicator of a second latent trait being measured by the items or due to the presence of very similar items the responses of which would then inflate the scores, altering the properties of the measure (Yang & Kao, 2014). When both the unidimensionality and the local independence assumptions are met, the person ability parameters estimated in the model are not test-dependent and the item parameters are not sample-dependent.

Item Response Theory uses a mathematical model to predict the probability of a subject answering an item correctly, depending on the personal ability parameter and the estimated difficulty of the item. The relationship between item and personal parameters is usually graphed in an item characteristic curve (ICC) that has a shape similar to an 'S', displaying the ability level or latent trait level on the X-axis and the probability of a correct or positive response on the Y-axis.

Considering that the data to be analyzed in this study was composed of dichotomic responses to the items, the IRT analyses were conducted using the Rasch model. This model works with answers in a dichotomic format, where 1 indicates a correct response and 0 an incorrect response. Both item difficulty and person ability are modeled on the same scale, so a single continuum is used to represent the relationship between them. In the Rasch model, the probability of a respondent with ability level θ_i correctly answering an item with difficulty b_j is calculated using a logistic function with the following equation (Meyer, 2014):

$$P_{ij}(\theta) = \frac{\exp(\theta_i - b_j)}{1 + \exp(\theta_i - b_j)}$$

This probability is determined by the item's difficulty and the respondent's ability or level of the latent trait and can be graphically displayed by the item characteristic curve in Figure 3.17. The curve shows how the probability monotonically increases as the ability value rises. For higher values of the latent trait, the function tends asymptotically towards 1, showing an increasing probability of answering the item correctly. For lower parameters of the latent ability, the function tends to 0, revealing a lower probability of correctly answering the item.

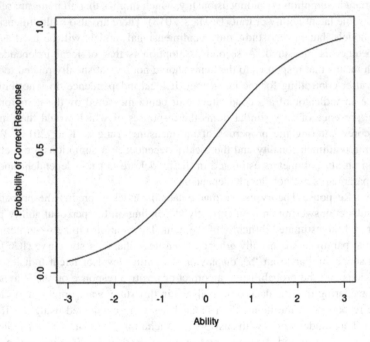

Figure 3.17 Example of an Item Characteristic Curve (ICC)

In other words, if the item difficulty is greater than the respondent's ability parameter, a correct response is unlikely to be produced. On the contrary, if the ability parameter is higher than the item difficulty, the probability of a correct answer exceeds 50%. If item difficulty and latent trait parameters are the same, the chances of the subject answering correctly are 50%.

In the graph, the item difficulty parameter determines the location of the item characteristic curve. For difficult items, the entire ICC is shifted to the right on the X-axis and for items with low difficulty values, the curve is shifted to the left. Figure 3.18 displays the ICCs for three items with different difficulty values. The curve on the left corresponds to a low difficulty item for which the probability of a correct response is greater also for lower ability parameters. Conversely, the curve on the rightmost illustrates a difficult item that lowers the probability of success for any given ability value.

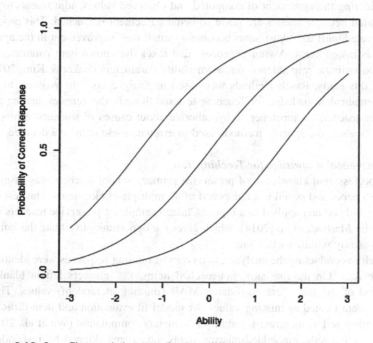

Figure 3.18 Item Characteristic curves for three different item difficulties

An additional distinguishing property of the Rasch model is specific objectivity (Rasch, 1977; Wu & Adams, 2007), which refers to the invariant comparisons that can be made between persons and between items. Because no additional parameters, such as item discrimination, are included in the Rasch model, the difference in the ability parameters of two subjects is independent of the items they answered and the difference in the calculated item difficulties is independent of the respondents' abilities.

For the application of the Rasch model, as has already been mentioned, two parameters have to be estimated, i.e., the item parameter, representing the difficulties of the items, and the person parameter, accounting for the level of the ability or latent trait. To find the item characteristic curve that best fits the proportions of correct response observed in the data, the maximum likelihood estimation (MLE) procedure is used. Initially, a priori values are set for the item parameters and used to compute the probability of a correct response at each ability level. Considering the agreement of computed and observed values, adjustments to the estimated item parameters are made so better agreement is reached. This process is repeated until the adjustments become so small that improvement in the agreement is insignificant. A similar process that takes the known item parameters is used to estimate respondents' unknown ability parameters (Baker & Kim, 2017).

In this study, Rasch methods were used to analyze both the answers to the Mathematical Knowledge for Teaching test and those to the items evaluating preservice teachers' competence to hypothesize about causes of students' errors. In the following, the specific methods used to evaluate these items are detailed.

Mathematical Knowledge for Teaching Test
The professional knowledge of preservice primary school teachers was assessed using a paper and pencil test composed of 45 multiple-choice items. This test had been validated and applied to a bigger Chilean sample of preservice teachers in a study by Martínez et al. (2014), which gives the opportunity to situate the sample of this study within a wider one.

Before conducting the analyses, two types of missing responses were identified in the data. On the one side, not-reached items, i.e., answers left in blank at the end of the test, were considered MAR (missing at random) values. These items were treated as missing values for model fit evaluation and item difficulty estimation and as incorrect for ability parameters computation (Wu et al., 2016). On the other side, embedded-missing items, i.e., items skipped by respondents that are followed by other answered items, were considered MNAR (missing not at random) values because it is assumed that examinees would intentionally skip items when they do not have sufficient knowledge to judge the correctness of their intended response and, therefore, the probability of missingness is related to the measured variable. Thus, this kind of missing values was treated as incorrect in all analyses. Altogether, missingness was observed in a very small proportion, only 4,1% of the data set. According to the findings of Waterbury's (2019) simulation study, bias introduced by missing data in item difficulties estimation can "vary dramatically across the proportion of missing responses factor, and also the data mechanism factor" (p. 160). The results of his study suggested that whereas bias in item difficulties is negligible when missing values are MCAR or MAR, it is severe when they are MNAR. Yet, this severity decreased dramatically when the proportion of missing responses was 0.2 instead of 0.5.

The IRT analyses were conducted using the TAM package for the R software (Robitzsch, Kiefer & Wu, 2017). Results from the application of the Rasch model suggested a good fit. Table 3.5 displays the main indices. The item difficulty mean of the 45 test items is 0.1 (SD = 1.04007), with a range of difficulty levels from −2.057 to 2.066. Alternatively, the person ability parameters show a mean of −0.09 (SD = 0.86068), revealing that the items were slightly difficult for the examinees. The lowest person parameter is −2.064 and the highest is 2.288.

The item discrimination mean of the items is 0.347 (SD = 0.09687), where the smallest is 0.037 and the greatest is 0.503. Despite the low discriminating values of some items and considering that they showed at the same time good weighted mean square (MNSQ) values, they were included in further analyses in order to make comparisons with the national sample. The weighted MNSQ (Infit) of all items were within the 0.8 to 1.2 range suggested by Linacre (2002). The mean is 0.997 (SD = 0.06722), ranging from a lowest value of 0.888 to a highest value of 1.187. The expected a posteriori/plausible values or EAP/PV reliability provides a value of sample reliability that, although not the same, can be interpreted similarly to Cronbach's Alpha in Classical Test Theory. Therefore, indices over 0.7 are interpreted as supporting reliability (Neumann, Neumann, & Nehm, 2011). The Rasch analysis produced an EAP/PV reliability value of 0.825, confirming there is sufficient reliability.

Table 3.5 Mathematical knowledge for teaching test item characteristics

Item	Correct response percentage	Item discrimination	Item parameter	Infit	Infit t-value
Item 1	0.616	0.478	−0.530	0.906	−1.92
Item 2	0.658	0.503	−0.735	0.888	−2.04
Item 3	0.671	0.501	−0.798	0.897	−1.80
Item 4	0.322	0.459	0.835	0.927	−1.19
Item 5	0.678	0.384	−0.831	0.972	−0.46
Item 6	0.596	0.489	−0.431	0.911	−1.89
Item 7	0.778	0.153	−1.405	1.091	1.05

(continued)

Table 3.5 (continued)

Item	Correct response percentage	Item discrimination	Item parameter	Infit	Infit t-value
Item 8	0.774	0.417	−1.378	0.929	−0.82
Item 9	0.686	0.316	−0.882	1.023	0.39
Item 10	0.294	0.394	0.991	0.968	−0.47
Item 11	0.586	0.426	−0.392	0.957	−0.91
Item 12	0.136	0.360	2.066	0.946	−0.37
Item 13	0.468	0.352	0.148	1.020	0.47
Item 14	0.627	0.460	−0.588	0.932	−1.34
Item 15	0.729	0.415	−1.103	0.944	−0.77
Item 16	0.247	0.450	1.256	0.925	−0.94
Item 17	0.866	0.232	−2.057	1.017	0.18
Item 18	0.500	0.376	0.004	1.001	0.04
Item 19	0.511	0.318	−0.050	1.041	0.95
Item 20	0.432	0.422	0.312	0.965	−0.76
Item 21	0.155	0.254	1.894	1.029	0.28
Item 22	0.201	0.216	1.549	1.073	0.77
Item 23	0.165	0.326	1.815	0.974	−0.19
Item 24	0.283	0.308	1.052	1.028	0.42
Item 25	0.583	0.354	−0.377	1.007	0.17
Item 26	0.745	0.037	−1.200	1.187	2.34
Item 27	0.744	0.411	−1.193	0.952	−0.61
Item 28	0.197	0.340	1.568	0.988	−0.09
Item 29	0.208	0.408	1.505	0.924	−0.78
Item 30	0.715	0.367	−1.025	0.965	−0.48
Item 31	0.603	0.340	−0.461	1.012	0.26
Item 32	0.518	0.355	−0.075	1.008	0.19
Item 33	0.376	0.370	0.584	0.998	−0.02
Item 34	0.394	0.298	0.498	1.060	1.14
Item 35	0.459	0.307	0.198	1.052	1.11
Item 36	0.367	0.313	0.615	1.041	0.73

(continued)

Table 3.5 (continued)

Item	Correct response percentage	Item discrimination	Item parameter	Infit	Infit t-value
Item 37	0.394	0.327	0.498	1.025	0.49
Item 38	0.408	0.239	0.425	1.091	1.76
Item 39	0.421	0.157	0.365	1.161	3.10
Item 40	0.462	0.221	0.174	1.107	2.23
Item 41	0.472	0.302	0.120	1.052	1.11
Item 42	0.217	0.399	1.435	0.944	−0.57
Item 43	0.148	0.394	1.934	0.911	−0.65
Item 44	0.620	0.375	−0.572	0.992	−0.13
Item 45	0.742	0.290	−1.199	1.030	0.40

Taking the estimated item difficulties as a base, ability parameters were also computed with the TAM package for the R software. Before standardizing ability parameters to a mean of zero and a standard deviation of one and scaling them to a mean of 50 and a standard deviation of 10 for ease of further interpretation of the data, the calculated weighted likelihood estimates (WLE) were used to compare the results of participants in this study to those of the participants in the Chilean study mentioned above (Martínez et al., 2014).

In the study by Martínez et al. (2014), a bigger sample of Chilean preservice teachers answered this MKT test (N = 421). The mean obtained by preservice primary teachers in that study was 0.323 (SD = 0.157), whereas in the present study it was 0.474 (SD = 0.161). The median was also considerably higher in the present study (Mdn = 0.478) than in Martínez et al.'s (2014) study (Mdn = 0.311). The range of ability parameters was smaller in this study (0.73) than in Martínez et al.'s (2014) study (0.911). It is also worth noticing that percentile 75 in Martínez et al.'s (2014) sample is located on a lower ability parameter (0.400) than the percentile 50 in the present study (0.478), as can be seen on the box plots in Figure 3.19. Altogether, participants in the present study exhibited higher ability parameters in the mathematical knowledge for teaching test than preservice teachers who took part in the study by Martínez et al. (2014).

Competence to Hypothesize about Causes of Students' Error
Preservice teachers' competence to make hypotheses about the causes of students' errors was measured in this study using closed items as described in section 3.1.2 and illustrated in Figure 3.11. Respondents were faced with a list of causes for

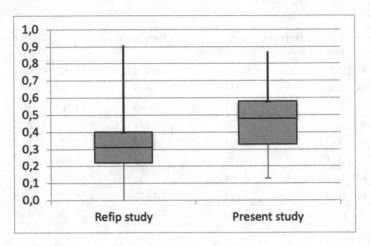

Figure 3.19 MKT ability parameters comparison between the present study and the Refip study

the particular error and they had to decide if each of them was 'a possible error' or 'not a possible error'.

The statements of possible and not possible causes for each of the four error-situation tasks were developed based on the relevant literature. In order to decide if each of the given causes should be interpreted as a possible or as a not possible cause, the items were rated by a group of experts, following the guidelines given by the expert ratings conducted within the TEDS-FU study, albeit in a smaller scale (Kaiser et al., 2015). The group of experts was composed both of scholars from the mathematics education field, with a focus on empirical studies, and primary school education and of experienced teachers. In total, eighteen experts from Chile and Germany evaluated the items. The items were accepted to be included if at least a 65% of agreement was reached. Items with a weaker agreement in the expert rating were dropped out.

The analyses of the items under the unidimensional Rasch model, were conducted using the TAM package for the R software (Robitzsch et al., 2017).

The data set was explored prior to the analyses and two different missing values mechanisms were identified. The testing design of the study, in which participants answered randomly assigned booklets at the pre- and post-test, produced missing-by-design missing responses. These are considered MCAR (missing completely at random) values, as missingness is unrelated to any other variables and the probability of missingness is the same for all examinees. These values were treated as missing in all analyses. Additionally, embedded-missing items were identified. These were treated as MNAR values, as it can be assumed that respondents intentionally left the items in blank when they were unsure of the response and, therefore, the missingness is related to the competence being measured. MNAR values were treated as incorrect in all analyses.

Parameter estimation took place in two phases. First, information on the items was extracted, including the estimation of difficulty parameters. This was done using all available data of preservice primary school teachers who answered the pre-test. In a second phase, ability parameters were calculated for all participants in the study.

To decide if the set of items of the four error tasks had a similar difficulty level, respondents' raw scores on the two error-tasks answered at each testing-time were compared. Analyses yielded significant correlations within each booklet, suggesting that the four sets of items had similar properties and thus can be further analyzed together.

Before doing the definitive item parameters estimation, an additional item selection was conducted considering preliminary item discrimination and Infit indices. Items showing discrimination values smaller than 0.25 and/or Infit values out of the 0.8–1.2 range recommended by Linacre (2002) were excluded from further analyses. Table 3.6 shows the main parameters of the analysis performed with the remaining items. The mean item-difficulty is −0.704 (SD = 1.35535), for which the easiest item has a difficulty parameter of −3.248 and the hardest one a value of 2.078. Regarding item discrimination, results indicate a mean of 0.423 (SD = 0.101786) with a range from 0.213 to 0.603. The mean of the weighted MNSQ (Infit) values is 0.997 (SD = 0.06667), with a lowest value of 0.875 and a highest value of 1.209, all within the range suggested by Linacre (2002).

The reliability of the scale can be evaluated using the EAP/PV reliability values. For this set of items assessing preservice teachers' competence to hypothesize about causes of students' errors, results yielded an EAP/PV reliability of 0.648. Although this value is slightly lower than the traditional 0.7 threshold, the items were used for further analysis, taking into consideration the complexity of the competence that is being assessed and the fact that the survey was developed especially for this study on a rather small scale, as a first approach to developing such an assessment for preservice primary school teachers.

In a second phase, the person ability parameters were calculated. As it was done to calculate item parameters, maximum likelihood procedures were used to estimate respondents' ability parameters. In particular, the weighted likelihood

Table 3.6 Item characteristics for the items evaluating the competence to hypothesize about causes of students' errors

Item	Response percentage	Item discrimination	Item parameter	Infit	Infit t-value
Causes Subtraction inversion 1	0.885	0.355	−2.354	1.001	0.05
Causes Subtraction inversion 2	0.894	0.252	−2.451	1.048	0.35
Causes Subtraction inversion 3	0.688	0.602	−0.919	0.875	−1.90
Causes Subtraction inversion 4	0.236	0.359	1.393	1.050	0.61
Causes Subtraction inversion 5	0.867	0.253	−2.169	1.068	0.53
Causes Subtraction inversion 6	0.566	0.501	−0.300	0.982	−0.34
Causes Subtraction inversion 7	0.603	0.426	−0.470	1.049	0.87
Causes Subtraction inversion 8	0.817	0.519	−1.741	0.914	−0.81
Causes Addition 1	0.647	0.442	−0.693	1.042	0.65
Causes Addition 2	0.440	0.458	0.313	1.022	0.40
Causes Addition 3	0.535	0.603	−0.127	0.913	−1.64
Causes Addition 4	0.688	0.488	−0.918	0.979	−0.27
Causes Addition 5	0.917	0.271	−2.752	1.027	0.20

(continued)

Table 3.6 (continued)

Item	Response percentage	Item discrimination	Item parameter	Infit	Infit t-value
Causes Numbers Transcoding 1	0.937	0.489	−3.072	0.879	−0.29
Causes Numbers transcoding 2	0.441	0.406	0.247	1.007	0.11
Causes Numbers Transcoding 3	0.432	0.444	0.289	0.994	−0.09
Causes Numbers Transcoding 4	0.541	0.415	−0.214	1.037	0.53
Causes Numbers Transcoding 5	0.396	0.514	0.460	0.931	−0.96
Causes Numbers Transcoding 6	0.946	0.309	−3.248	0.969	0.02
Causes Numbers Transcoding 7	0.225	0.450	1.385	0.947	−0.40
Causes Subtraction with Zero 1	0.840	0.414	−1.944	0.981	−0.05
Causes Subtraction with Zero 2	0.713	0.476	−1.083	0.986	−0.10
Causes Subtraction with Zero 3	0.670	0.213	−0.851	1.209	1.95
Causes Subtraction with Zero 4	0.138	0.288	2.078	1.043	0.26
Causes Subtraction with Zero 5	0.447	0.502	0.235	0.977	−0.29
Causes Subtraction with Zero 6	0.606	0.491	−0.526	0.973	−0.30
Causes Subtraction with Zero 7	0.362	0.382	0.650	1.059	0.66
Causes Subtraction with Zero 8	0.688	0.521	−0.934	0.948	−0.47

estimates (WLE) provided by R package TAM were used. For this procedure, the previously estimated item difficulty parameters were used to compute the probability of correct responses for each participant at both testing times. Results show a mean ability parameter of -0.062 (SD $= 1.01047$) and a range from -3.670 to 1.869 in the first testing-time and a mean of 0.199 (SD $= 0.99946$) with a lowest parameter of -1.721 and a highest parameter of 3.496 in the second testing-time. This shows that the items were easier for the participants after the seminar sequence.

To facilitate estimates interpretation, ability parameters were converted into T-scores. To do this, parameters were first standardized to a distribution with a mean of zero and a standard deviation of one. The standardization of the ability parameters obtained by participants in the first testing-time (t_1) was performed using the calculated mean ($\overline{x}_{t_1}$) and standard deviation (S_{t_1}):

$$z_{t_1} = \frac{x_{t_1} - \overline{x}_{t_1}}{S_{t_1}}$$

To avoid identical distribution curves, means and standard deviations between pre- and post-test results and so allow for comparison and evaluation of the effect of the university seminar sequence on the development of preservice teachers' competence to hypothesize about causes of students' errors, the standardization of the ability parameters of the post-test was conducted using the first measurement point mean ($\overline{x}_{t_1}$) and standard deviation (S_{t_1}). So, the standardization of second testing-time ability parameters was conducted using the following formula:

$$z_{t_2} = \frac{x_{t_2} - \overline{x}_{t_1}}{S_{t_1}}$$

After this, the standardized scores were converted into T-scores by scaling them to a mean of 50 and a standard deviation of 10. The formula used to perform the calculation for the first testing-time data is displayed below. The one used at the second testing-time is equivalent.

$$T_{t_1} = 10 * z_{t_1} + 50$$

3.3.2.2 Latent Class Analysis

Latent Class Analysis (LCA) is a mixture model used to identify subgroups in empirical data. It postulates that there is an unobservable, or latent, categorical variable that allows categorizing the population into distinct latent classes (Lanza & Rhoades, 2013). This latent variable is measured indirectly by means of a set of observed variables, also called indicators, which are subject to measurement error (Collins & Lanza, 2010). There are two key aspects that distinguish LCA from other latent variable models, namely the way in which the indicators are treated and the nature of the latent variable. In LCA, indicators are treated as categorical observed variables, as opposed to being considered continuous in other models. Additionally, and in contrast to other models with a continuous latent variable, in LCA the latent variable is categorical, i.e., it focuses on qualitative differences among groups (Collins & Lanza, 2010). In other words, LCA is concerned with the characterization of the heterogeneity on the observed indicators in the studied population by identifying a number of homogeneous subgroups that are also distinct among them. These groups, derived from statistical analyses of response patterns, are called latent classes because the categorical variable forming the groups is not measured directly (Feingold, Tiberio & Capaldi, 2014).

In this sense, LCA is considered a person-oriented approach, according to the differentiation made by Bergman and Magnusson (1997) between variable-oriented and person-oriented approaches to statistical analyses in social sciences research. While variable-oriented approaches focus on relations between variables and assume that these can be applied to all individuals in the same way, person-oriented approaches put their emphasis on the individual, accounting for the heterogeneity of patterns in the nature of individual differences (Bergman & Magnusson, 1997; Collins & Lanza, 2010).

LCA allows the organization of a number of observed variables into homogeneous subgroups that are meaningful for the research question, especially when the size and complexity of the collected indicators do not permit identifying such patterns by simple inspection of the data (Collins & Lanza, 2010). The latent classes are assumed to completely explain the dependencies between the observed variables. In other words, local independence is assumed in that within each latent class, every manifest variable is statistically independent of every other manifest variable (Collins & Lanza, 2010). However, in the complete data set, the observed variables are not independent because the relations among them are precisely what is explained by the latent classes.

The LCA model characterizes the identified subgroups using two sets of estimated parameters (Hickendorff, Edelsbrunner, McMullen, Schneider, & Trezise, 2018). The first set, the estimated latent class parameters, specify class prevalence, indicating the estimated proportion of individuals assigned to each class. Because the latent classes are mutually exclusive and exhaustive, these parameters sum to 100%. The second set, the estimated class-conditional parameters, are the item-response probabilities. They indicate the probability of each response to each item or observed variable conditional to membership in a given class. Careful examination and interpretation of these probabilities lead to an understanding of the essential characteristics of the latent classes and their labeling. Estimation of LCA parameters is done using iterative algorithms, usually the maximum likelihood criterion (Vermunt & Magidson, 2004).

As a whole, patterns of item-response probabilities should display two essential characteristics, homogeneity and latent class separation (Collins & Lanza, 2010). Homogeneity refers to the extent to which the same response pattern is observed within a latent class and therefore describes this class. When the homogeneity of a class is high, most item-response probabilities for that class are either close to 0 or to 1. Latent class separation, on the contrary, refers to the extent to which the patterns of item-response probabilities of the different observed variables are clearly distinct among latent classes. In other words, when a model shows high latent class separation, the response pattern that typifies one of the classes will have a very small probability of occurring on any other class. However, it is important to notice that perfect homogeneity and latent class separation are not to be expected in empirical data but to be used as a reference to evaluate latent class models and assist their interpretation (Collins & Lanza, 2010).

When both homogeneity and latent class separation are high in a model solution, the probabilities of membership of each individual into one latent class tend to be high for one class and very low for the remaining classes, leaving a small degree of classification uncertainty. The probability of an individual being assigned into one class, given their pattern of responses, is called the classification probability or posterior probability (Collins & Lanza, 2010; Hickendorff et al., 2018). When large probabilities of individuals belonging to one class and low probabilities of belonging to the other classes are obtained, it can be said that there is a high degree of classification certainty. On the contrary, when for many individuals the posterior probabilities are far from 1, there is a high degree of classification uncertainty. Nevertheless, in empirical data sets, there are usually at least a few individuals whose response patterns show a high degree of classification uncertainty (Collins & Lanza, 2010).

One way of summarizing the degree of classification uncertainty is the mean posterior probability of each latent class. The mean and variability are calculated considering the probabilities of all individuals whose largest classification probability is at that same latent class, thus providing a measure of classification certainty for each latent class (Collins & Lanza, 2010). Another approach is based on providing a measure of entropy, for which larger values are an indication of less classification error. Because the entropy measure is a weighted mean of the classification probabilities of all individuals, it is a value between 0 and 1. It is worth noticing that even when entropy values are considerably close to 1, there may be individuals who show high classification uncertainty (Collins & Lanza, 2010).

Model Selection
The selection of an LCA model is a critical issue because the calculation and interpretation of all the model parameters are based on the extracted number of classes. Thus, model selection is done under consideration of a variety of evaluation tools that include the assessment of absolute fit of a certain model and relative fit of multiple models (Collins & Lanza, 2010; Lanza & Rhoades, 2013). In addition, parsimony and interpretability need to be considered. According to parsimony, a model should be as simple as possible, i.e., have as few parameters as possible to represent the data. Interpretability stands for the extent to which the model is understandable and each class can be interpreted under the lens of prior knowledge and experience in the field (Collins & Lanza, 2010).

The likelihood-ratio chi-square statistic is commonly used to measure absolute model fit in LCA. It tests if the data are adequately represented by the specified latent class model by establishing the null hypothesis that there is no significant difference between the expected and the observed cell counts in the corresponding contingency table. Therefore, a significant p-value would indicate that the data do not fit the LCA model. In other words, to confirm LCA model fit in absolute terms, the null hypothesis should not be rejected (Collins & Lanza, 2010). However, this statistic is useful when the sample size is large enough and the item count is small, so the degrees of freedom are not too large and sparseness does not become an issue (Lanza & Rhoades, 2013). This occurs to a very limited extent in the present study.

In addition, different statistics can be used to assess the relative model fit of one model compared to other solutions with more or fewer classes. A common approach for assessing relative fit is the bootstrap likelihood-ratio difference test (BLRT), in which based on a bootstrap generated sample for the smaller model, a test statistic is calculated and then compared to that of the empirically obtained data. A p-value for the test statistic calculated from the empirical data set can be obtained taking the test statistic from the generated data as a reference distribution. Thus, a significant value of this test indicates that the model with more latent classes fits the data better than the model with one class less (Collins & Lanza, 2010; Dziak, Lanza & Tan, 2014).

Tekle, Gudicha and Vermunt (2016) found that BLRT has a high power of detecting the true model when class separation is moderate (0.8) to high (0.9) irrespective of other conditions, such as sample size, number of indicator variables or number of classes. Moreover, they indicated that the power of the test also increases with sample size and it slightly lowers when class sizes are unequal. However, they indicate that even with moderate class separation, sample size does not need to be very high for the BLRT to have powerful capability of detecting the true model. Additionally, Dziak et al. (2014) stated that unequal class sizes and low class separation have a strong effect on the power of BLRT and that when the sample size is insufficient, BLRT is likely to identify too few classes.

Other approaches to evaluate relative model fit are based on information criteria like the Akaike Information Criterion (AIC), Bayesian Information Criterion (BIC) and adjusted BIC. Lower values suggest a better balance between model fit and parsimony (Collins & Lanza, 2010; Lanza & Rhoades, 2013). Dziak et al. (2014) point out that using BIC to choose between two models is likely to under-fit, i.e., to select an oversimplified model, if sample size is not large. On the other hand, they argue, AIC is likely to lead to the selection of an overfitted model. Thus, they advocate for the simultaneous consideration of both measures. Based on a different simulation study, Nylund, Asparouhov and Muthén (2007) claimed that the BIC value should be decisive for choosing the best model solution. Collins and Lanza (2010, p. 88) suggest that 'information criteria are likely to be more useful in ruling out models and narrowing down the set of plausible options than in pointing unambiguously to a single best model'.

In this study, latent classes were extracted from preservice teachers' answers to the decision-making items. Latent class analysis permitted to find different types of preferences for dealing with students' errors among participants. In the following, the specific methods used to unravel these preference classes are detailed.

Preferences for Dealing with Students' Errors

In the survey, after the items about the causes of students' errors, preservice teachers were asked to suggest three alternatives to support the student in overcoming their error, as described in section 3.1.2 and shown in Figure 3.12. These open items were then coded using evaluative qualitative text analysis, as described in section 3.3.1.

Each of the three alternatives to dealing with the student's error was coded with one level (strong evidence of the category, some evidence of the presence of the category, no evidence of the category and unable to classify) for each of the five categories included in the analysis (conceptual understanding, procedural understanding, targeted strategies, active-learner approach and teacher-directed approach). To simplify the analysis and make the extraction of latent classes possible, the three answers or suggestions from each individual were taken together and summarized according to the frequency of the level codes. If all three answers were coded with either 'no evidence' or 'unable to classify', the individuals' answers were summarized as 'no evidence'; if within the three answers the highest level was 'some evidence', they were summarized as 'some evidence' if it appeared only once, and as 'repeated some evidence' if it emerged twice or three times. Similarly, if the highest level within the three answers was 'strong evidence', they were summarized as 'strong evidence' if this level code showed up once and as 'repeated strong evidence' if it occurred more than once. From these summarized levels, the latent classes were then extracted.

According to the testing design described in section 3.1.2, each participant answered two of the four error analysis tasks at each testing time. As a result, their answers received two sets of evaluative qualitative text analysis codes at each the pre- and post-test. The latent classes extracted with these codes were significantly related to the error analysis tasks, indicating that participants showed different preferences according to the error situations. By looking at the four error situations in retrospective, it can be seen that there is one task, the number transcoding error, that actually has a different nature than the others. Therefore, further analyses were conducted considering only one task per measurement point, namely both subtraction tasks.

The first step in LCA is conducting a series of analyses to fit models with an increasing number of classes. In this case, five models starting with a 2-class model were estimated using the Mplus 7 software (version 1.4 (1); Muthén & Muthén, 1998–2015), which applied maximum likelihood estimation to estimate the models. The relevant measures for model selection, namely AIC, BIC, aBIC and BLRT, and other relevant measures like entropy, were also included in the analyses with the software.

Regarding absolute fit indices, the chi-square test is non-significant for all five evaluated models. Thus, all models are adequately represented by the data. Additionally, entropy values are high for the five models, ranging from 0.83 to 0.93, and the classification probabilities are above 0.8 for the models with four, five and six classes and above 0.9 for models with 2 and 3 classes.

Relative model fit measures point mainly to the three-class and the four-class solutions. As shown in Table 3.7, the AIC shows very similar values for the four-, five- and six-class solutions, which are also lower than those of the two- and three-class solutions. The adjusted BIC shows the smallest value in the four-class solution. Alternatively, the BIC for the three-class model was smaller than the BIC found for any of the other models, suggesting that this model should be selected. The BLRT, though, was statistically significant at the 0.1 level for the four- but not for the five-class model, suggesting that the model with four rather than the one with three classes should be preferred. As suggested by Collins and Lanza (2010), these criteria were taken as hints to narrow down the number of possible solutions.

Besides the measures of absolute and relative fit, the interpretability of the results from a theoretical perspective is crucial for the selection of the model. The three-class model allowed for a better interpretation than all the other models. Moreover, the selection of the three-class model was favored for parsimony reasons and because it had very high classification probabilities, which are important for the use of the results for further analyses.

Table 3.7 Fit indices of different class solutions for the preferences for dealing with student errors. Note: lowest values in bold.

Number of classes	Entropy	AIC	BIC	aBIC	BLRT p-value
2 classes	0.892	3251	3397	3267	<0.001
3 classes	0.830	3143	**3364**	3168	<0.001
4 classes	0.830	**3132**	3428	**3165**	<0.001
5 classes	0.844	**3131**	3502	3173	0.04
6 classes	0.927	**3131**	3577	3181	0.03

The three classes were considerably different and clearly interpretable. The graph in Figure 3.20 shows the probabilities for the members of each class of demonstrating at least 'repeated some evidence' for the category displayed on the x-axis. That is, that within the three pedagogical suggestions for dealing with the student's error, participants' answers were summarized as showing either 'repeated some evidence', 'strong evidence' or 'repeated strong evidence'.

The first preference class, in which the biggest proportion of participants were classified, can be labeled as the 'instructivist preferences class'. This class of preservice teachers has the highest probability of providing evidence of a teacher-directed approach in their pedagogical suggestions for dealing with the student's error. This means that their answers focus on what the teacher would do in the situation and usually use verbs such as 'tell', 'show' or 'explain'. Their answers are likely to be targeted to the student's current understandings, although to a lesser extent than for participants in the second class. Finally, the probabilities

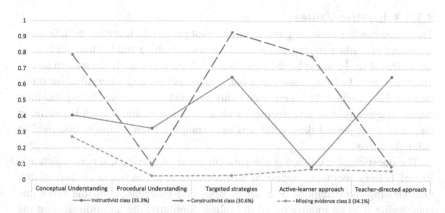

Figure 3.20 Probability of displaying at least repeatedly some evidence of the category for the three-latent-class model of the preferences for dealing with student errors

that the answers of preservice teachers in this class aim at promoting students' conceptual or procedural understanding are very similar.

The second, and smallest class, was composed of preservice teachers with a very distinguishable profile, named 'constructivist preference class' hereafter. They show very high probabilities of focusing on promoting a conceptual understanding of mathematics in order to support students to overcome their errors. On the contrary, it is very unlikely that their pedagogical suggestions pay attention to procedural understanding. Furthermore, they have the highest probability (significantly higher than class 1) of suggesting a strategy that is targeted to the specific error in that particular situation, accounting for the student's knowledge and understanding. Moreover, participants in this class have a very high probability of providing evidence of an active-learner approach and a very low probability of exhibiting a teacher-directed approach. Their strategies focus on what the student would do, think or answer.

Finally, the model extracted a third class showing very low probabilities of providing evidence for any of the five categories. In this class, only a low probability of focusing on conceptual understanding can be distinguished. This class accounts for those preservice teachers who provided very scant answers for the items to dealing with students' errors.

3.3.2.3 Ipsative Values

The questionnaires collecting information about preservice teachers' beliefs about the nature of mathematics and about mathematics teaching and learning stem from the TEDS-M study, as explained in section 3.1.2. As it is usually the case in instruments designed to gather data about individuals' beliefs and other affective traits, they use Likert scales. In Likert scale items, respondents are asked to express their degree of agreement or disagreement with a statement within a provided scale. In this case, a six-point Likert scale was used, ranging from 'strongly disagree' to 'strongly agree'.

The questionnaire collecting information about preservice teachers' epistemological beliefs about the nature of mathematics was composed of two scales. One viewing the discipline as a set of rules and procedures and the other one considering it as an inquiry-based discipline. Similarly, the questionnaire on the beliefs about mathematics teaching and learning was organized into two scales, representing student-centered and teacher-centered approaches.

To confirm the score reliabilities of these scales, i.e., their internal consistency, Cronbach's alphas were calculated. Cronbach alpha "is a measure of the internal consistency among the *items*" (Cohen, Manion & Morrison, 2007, p. 506) and therefore suggests if a set of items are measure the same construct. Thus, the alpha coefficient was calculated for each of the four scales at each testing time. The obtained reliabilities are displayed in Table 3.8 and indicate that all scales are reliable (0.7–0.79), highly reliable (0.8–0.9) or very highly reliable (>0.9), according to Cohen et al. (2007) suggestions.

Table 3.8 Reliability coefficients for beliefs scales

Scale	Testing time	Cronbach's Alpha
Beliefs about the nature of mathematics		
Rules and Procedures	Pre-test	.776
	Post-test	.831
Process of inquiry	Pre-test	.914
	Post-test	.792
Beliefs about the learning of mathematics		
Teacher-centered	Pre-test	.837
	Post-test	.758
Active learning	Pre-test	.913
	Post-test	.877

A common difficulty with this type of questionnaire is bias introduced by response styles. Tendencies to choose categories at the positive end of the scale, at the negative end, at the midpoint or at either extreme regardless of item content, usually varies among respondents (Weijters, Geuens & Schillewaert, 2010). One way to address this issue is by using ipsative values, as it has been done in studies such as TALIS (OECD, 2009) or TEDS-M (Felbrich, Schmotz & Kaiser, 2010) for the evaluation of their beliefs scales. The use of ipsative scores allows to express the relative preference of individual responses between two (or more) different views and thus to reduce the effect of bias introduced by response styles (OECD, 2009). To obtain an ipsative score for each respondent, their mean across the scores of both scales is subtracted from the mean across the items in the corresponding scale. In this way, the scores are corrected for the tendency to agree to any item. This also results in the mean of each pair of ipsative values to be zero and provides for each examinee a score that indicates their relative position on one scale in relation to the other. Positive values denote that the referred belief scale, corrected for the response tendency, is stronger agreed to in relation to the other scale (OECD, 2009; Felbrich et al., 2010).

In the present study, ipsative scores were calculated for both beliefs categories. For the beliefs about the nature of mathematics, preservice teachers showed a higher endorsement to the scale of mathematics as a process of inquiry than to mathematics as a set of rules and procedures in both testing times, as can be seen in Table 3.9. Similarly, in the category of beliefs about the teaching and learning of mathematics, preservice teachers agreed stronger with the active learner scale than with the teacher-centered scale both in the pre- and post-test.

In further analyses, ipsative scores of the scales closely related to constructivism, i.e., mathematics as a process of inquiry and active learning of mathematics, will be used to explore correlations with other variables. This procedure can be used because there is a direct relation of the values of both scales in each category, which means that a respondent's positive endorsement of one scale is strictly related to a rejection of the other scale.

3.3.2.4 Hypotheses Testing

In order to examine the hypotheses detailed in chapter 2 and study how preservice primary teachers observed diagnostic competence in error situations and its development are related to a set of other variables, various statistical analyses were conducted using the software SPSS Statistics, version 25.0.0.1. The characteristics of the variables under study are critical to the selection of the procedures for the analyses. Therefore, the procedures used to explore the relationships between the competence to hypothesize about causes of students' errors and the

Table 3.9 Ipsative scores means and standard deviation of beliefs items

Scale	Testing time	Mean	SD
Beliefs about the nature of mathematics			
Rules and Procedures	Pre-test	−0.46	0.59
	Post-test	−0.67	0.60
Process of inquiry	Pre-test	0.46	0.59
	Post-test	0.67	0.60
Beliefs about the learning of mathematics			
Teacher-centered	Pre-test	−1.30	0.62
	Post-test	−1.31	0.61
Active learning	Pre-test	1.30	0.62
	Post-test	1.31	0.61

independent variables differ from those used to study the relationships between the preferences for dealing with students' errors and the independent variables. The former is a continuous variable and the latter a categorical variable. In addition, the characteristics of each of the independent variables need to be considered to select the appropriate procedures. This led to the use of independent-samples t-tests and Pearson correlations to explore the relationship of the competence to hypothesize about causes of students' errors with several independent variables. For the study of the relationships between the preferences for dealing with students' errors and the independent variables, analyses of variances (ANOVA) and Chi-square distributions were used. Each of these procedures is presented next.

Independent-samples t-tests were used to explore the relations with the dependent variable when the independent variable was categorical and had two groups. The null hypothesis for these tests is that the means of both groups are equal. Thus, it is usually intended to reject the null hypothesis and accept the alternative hypothesis, i.e., that there is a significant difference between the means of both groups. A significance level, also called alpha, is set in advance commonly at the 0.05 level, which narrows down the probability of occurring in the sample if there was no relationship in the population to a 5% (Cohen et al., 2007; Mertens, Pugliese & Recker, 2017; Muijs, 2004). In other words, the independent-samples t-tests were used to examine whether the population means of the two groups are different and that those differences are not just observed by chance in the sample means. This test requires that the dependent variable follows an approximately normal distribution within each group; yet, it is recognized as a robust test that bears some deviation away from normality. Additionally, this test assumes homogeneity of variances, i.e., that the variances of the two groups are equal in the population. This is measured using Levene's Test of Equality of Variances, which is run by default in SPSS when calculating the independent t-test. Whenever this assumption is not met and a different procedure should be used, this will be made explicit and the alternative procedure clarified.

For the t-tests that yielded statistically significant differences between the means of the two groups and thus suggested significant relationships between the competence to hypothesize about causes of students' errors and one of the independent variables, a standardized difference index of effect size was computed. In particular, Cohen's d statistic (Cohen, 1988) was calculated using the following formula:

$$d = \frac{M_1 - M_2}{\sqrt{\frac{s_1^2 + s_2^2}{2}}}$$

Where d = Cohen's d statistic, M1 and M2 are the means for the two groups being compared and an estimate of the population standard deviation is used as denominator, which is obtained by calculating the square root of the mean of the two squared standard deviations (Cohen, 1988; Hatcher, 2013). The obtained effect sizes are commonly interpreted as small ($d = 0.2$), medium ($d = 0.5$) and large ($d = 0.8$), following Cohen's (1988) criteria. However, these values should be used with caution, not as fixed cut-off points, and serve only as reference when previous research does not provide sufficient information against to which comparisons can be made (Cohen et al., 2007; Hatcher, 2013; Lakens, 2013).

Alternatively, to evaluate the association between the competence to hypothesize about causes of students' errors and continuous independent variables, *Pearson correlations* were calculated. A Pearson correlation coefficient (r) indicates the strength and the direction of a linear relation between two metric variables. It can range from -1 to $+1$. The sign of the coefficient indicates the direction of the relationship, i.e. if there is a positive or a negative relationship between the variables. A coefficient of zero indicates no relationship between both variables and absolute values closer to 1 indicate a stronger association between the variables. Criteria to interpret the coefficients, suggest that $r = \pm.10$ signifies a small effect, $r = \pm.30$ denotes a medium effect, and $r = \pm.50$ represents a large effect (Hatcher, 2013).

To investigate the relationships between the preferences for dealing with students' errors, a categorical variable with three groups, and continuous demographic variables, multiple *analysis of variance (ANOVA)* were used. The null hypothesis for the ANOVAs was that there is no significant difference among the means of the corresponding demographic variable among the three groups of preferences for dealing with students' errors. To test this null hypothesis, an *F* statistic and a corresponding *p*-value are calculated. The *F* statistic is the ratio of the between groups variance and the within group variance and it has a value of approximately 1.00 when there is no association between the independent and the dependent variable. On the contrary, when the differences can be explained by group membership, the *F* statistic is greater than 1.00. The related *p*-value indicates how likely it is that the differences found in the sample would occur if there were no differences between the groups in the population. As usual, a cut-off point of 0.05 is taken to determine whether the differences found are statistically significant. A significant *F*-statistic only indicates that there is a difference between at least two of the groups under comparison, but it does not show where the significance lies. To understand the nature of the difference, various post-hoc comparisons tests exist, which make individual comparisons between the groups of the categorical variable. Commonly, the Tukey HSD test is selected as the most appropriate because it provides the *p*-value for all possible pairwise comparisons and the confidence intervals for the mean difference for each comparison (Cohen et al., 2007; Mertens et al., 2017; Muijs, 2004).

As for the t-tests, the analysis of variance requires that there should be no significant outliers and the continuous variable should be approximately normally distributed for each category of the categorical variable. However, ANOVA is also a robust test in that it tolerates some violation of this assumption and can still provide valid results. When the dependent variable is departed from a normal distribution in one or more groups, the non-parametric Kruskal-Wallis test can be conducted (Sheskin, 2000). In addition, homogeneity of variances is assumed, i.e., the variances on the continuous variable should be equal across groups. In cases of violation of this assumption and unequal sample sizes of the groups, a recommended alternative test is Welch's (1951) method because of its' overall performance and computational ease (Jan & Shieh, 2014). Whenever an alternative test is used in this study, it will be made explicit.

To assess the strength of association between the variables when the ANOVA yielded a statistically significant result, partial eta squared (η_p^2) was computed. As an index of effect size, it represents the percentage of variance in the continuous variable accounted for by the categorical variable. For one-way ANOVA with one between-subjects factor, partial eta squared is always equal to eta squared and is computed according to the following formula:

$$\eta_p^2 = \frac{SS_B}{SS_B + SS_W}$$

Where SS_B is the sum of the squares between groups and SS_W is the sum of squares within groups (Hatcher, 2013). As with Cohen's d, it is recommended to interpret η_p^2 by comparing it to results of other studies. When this is not possible, commonly $\eta_p^2 = 0.01$ is considered as a small effect, $\eta_p^2 = 0.06$ is interpreted as a medium effect and $\eta_p^2 = 0.14$ as a large effect (Cohen, 1988; Cohen et al., 2007, Hatcher, 2013).

To study the relationships between the preferences for dealing with students' errors and categorical demographic variables, contingency tables were generated, and *chi-square tests of independence* were performed. A contingency table, also known as a crosstabulation, is a table displaying the frequencies for each combination of the categories of two or more nominal or ordinal variables. Additionally, the expected number of cases that should fall into each combination if there were no relationship between both variables are included. In other words, the distribution that is expected to occur by chance in the population, which is statistically calculated taking as a reference the percentage of either the variable on the rows or the variable on the columns of the whole. The null hypothesis for a chi-square (χ^2) test states that there is no statistical difference between the observed results and the expected cell counts. The chi-square test is used to determine if the difference between the actual distribution and that expected to occur by chance is statistically significant. It is computed using the following formula:

$$x^2 = \sum \frac{(O - E)^2}{E}$$

Where O are the observed frequencies and E the expected frequencies. If a statistically significant result is obtained (usually below the 0.05 level), it can be stated that the difference between the observed and expected frequencies is statistically significant and is not occurring simply by chance, i.e. the alternative hypothesis is supported (Cohen et al., 2007; Muijs, 2004). In such cases, knowing the strength of the association is necessary to complete the information. The most commonly used effect size test for the chi-square measure of independence is the Cramer's V test, which is calculated with the following formula

$$\sqrt{\frac{\chi^2}{n(k-1)}}$$

Where n is the total sample size and k the number of rows or columns in the contingency table, whichever is less (McHugh, 2013). Cramer's V ranges from 0 to 1, with larger values indicating a stronger association and because it is a form of a correlation, Cramer's V is interpreted following the same criteria (Kearney, 2017).

In addition, it is recommended to investigate where the significant differences are located. To do this, cells with the largest residuals can be found. In other words, those cells with greater differences between the observed and the expected values. Larger residuals make greater contributions to the obtained chi-square value. Therefore, locating them helps to understand the nature of the associations. (Sharpe, 2015). With the aid of adjusted standardized residuals and a Bonferroni adjustment of the critical value, observed values which are significantly discrepant with the expected values can be identified (García-Pérez & Núñez-Antón, 2003; MacDonald & Gardner, 2000).

Alternatively, to study the relationships between the development of both aspects of preservice primary teachers' diagnostic competence and the independent variables further statistical procedures were needed. In addition to chi-square tests of independence and ANOVAs, multiple regressions and analyses of covariance (ANCOVAs) were used.

To examine the connection between the changes in the competence to hypothesize about causes of students' errors from before to after their participation in the seminar sequence and the continuous independent variables, several *multiple regressions* were conducted. For each of them, the post-test ability parameter of the hypothesizing competence was taken as the dependent variable and both the pre-test ability parameter of the hypothesizing competence and the independent variable under study were included as predictors. Further details about multiple regressions can be found in section 3.3.2.6.

To evaluate if the changes in the competence to hypothesize about causes of students' errors can be associated to any of the categorical independent variables, several ANCOVAs were conducted. The analysis of covariance is an extension of the one-way ANOVA that incorporates a covariate variable. The covariate is a continuous independent variable, which is used to adjust the means of each group on the dependent variable. Because the covariate is influencing the dependent variable, it needs to be controlled for. In other words, including the covariate allows for a better evaluation of the differences between the means of two or more groups of the categorical independent variable, because it statistically controls for the effect of the covariate (Hatcher, 2013). Therefore, ANCOVA has been recognized as the most effective and powerful method for analyzing change on treatment studies (O'Connell et al., 2017). Moreover, for nonrandomized pretest-posttest research designs, the use of ANCOVA is recommended because it allows to adjust the means of the post-test results for differences among groups on the pre-test data (Dimitrov & Rumrill, 2003). In this study, the pre-test ability parameter of the competence to hypothesize about causes of students' errors was incorporated as a covariate in several ANCOVAs that took the post-test ability parameter as a dependent variable and individual demographic categorical variables, such as their teaching or private tutoring experience, as the independent grouping variable.

As with other tests, the strength of association between the variables was assessed when the ANCOVA yielded significant results. Partial eta squared was computed as a measure of effect size using the same formula as for the one-way ANOVAs, but taking the values adjusted by the covariate. The obtained value indicates the proportion of variance accounted for by the grouping variable after adjusting for the covariate and are interpreted using the same criteria as for the ANOVA (Hatcher, 2013).

Finally, to explore the changes in the preferences for dealing with student errors, different patterns of change were first identified and then used to study their relationship with various demographic variables. This was done using both analyses of variances and chi-square tests of independence. The former was used when the demographic variable was continuous and the latter when it was categorical.

Together, all these tests were used to examine the hypotheses of the present study. However, the presence of several hypothesis and statistical tests poses a statistical conflict that will be explained in the next section.

3.3.2.5 Problem of Multiple Testing

Many educational studies collect data on large samples and on many variables. When analyzing the data, typically several hypotheses are tested on the same sample. Therefore, several separate statistical tests are conducted to test the null hypothesis of no impact or no difference for each pair of variables, setting a statistical significance level of $\alpha = 0.05$ for each test. This means that the Type I error rate, i.e. the chance of erroneously finding a statistically significant result is 5 percent. However, when several statistical significance tests are performed, the error rate cumulates, leading to an inflated percentage chances of false rejection of the null hypotheses (Victor, Elsäßer, Hommel & Blettner, 2010). In other words, the probability of making a false conclusion increases dramatically with the number of conducted tests. For this reason, the family-wise error rate (FWER) has been defined as the probability that at least one null hypothesis will be rejected even if all of them are true (Kleinbaum & Klein, 2010). The FWER for independent tests is $1 - (1 - \alpha)^T$, where T is the number of independent tests. Hence, if the statistical significance p-values of ten tests are compared with an $\alpha = 0.05$, the probability of falsely rejecting at least one null hypothesis is of 40% (Victor et al., 2010). This rate increases to 64% if twenty tests are compared.

A very common way to control the FWER is the Bonferroni method (Cheng, Feng & Yi, 2017; Schochet, 2008; Victor et al., 2010). For ensuring that the FWER does not exceed a desired α (usually 5%), a new value is calculated by dividing α by the number of tests (T). The p-value of each test is compared to this new cut-off point to determine statistical significance. One disadvantage of this approach is that it is very conservative, making it more difficult to reject any null hypothesis and risking that results may be overlooked. An alternative modification of this method is the Holm-Bonferroni procedure, which provides increased statistical power (Schochet, 2008; Victor et al., 2010). It involves ordering the p-values of individual tests according to size. The smallest p-value is compared to the lowest limit, which is the same as in the Bonferroni approach, i.e. α/T. The second smallest p-value is compared to a value obtained by $\alpha/T - 1$. The third smallest p-value is compared to a less strict level, obtained by $\alpha/T - 2$. The procedure continues until the calculated value is exceeded for the first time. Null hypotheses with a p-value smaller than their corresponding limit are rejected and their alternative hypothesis are accepted.

In the initial analyses of the data of the present study, the significance levels of the tests used to examine the sub-hypotheses are given. Since the nature of this study is mainly exploratory, the results of the individual sub-hypothesis are first analyzed as isolated statements. These are only initial indications of associations and must in all cases be interpreted as such. They would need to be confirmed in further studies. Later, on the discussion of the results of the three overarching hypotheses, the problem of multiple testing is taken into consideration and the levels of significance are corrected using the Holm-Bonferroni method.

In addition to the individual testing of the sub-hypotheses, regression analyses were carried out in the present study to better understand the relationships implied in the hypotheses. Their methodological bases are briefly outlined below.

3.3.2.6 Multiple Regression Analysis

Multiple regression is an extension of simple linear regression, in which two or more independent variables (also called predictors or regressors) are used to predict a continuous dependent variable. The predictors can be continuous, dichotomic or categorical when appropriately coded. Considering a regression with k

predictors, x_1 to x_k, and a dependent variable y, regression coefficients are calculated for each predictor and a regression equation is formed (Fahrmeir, Kneib, Lang & Marx, 2013; Hatcher, 2013):

$$y = b_0 + b_1 x_1 + b_2 x_2 + b_3 x_3 + \cdots + b_k x_k$$

Where b_0 is the sample intercept, b_1 is the unstandardized slope coefficient for the first predictor x_1, b_2 is the unstandardized slope coefficient for the second predictor x_2 and so forth. Each b coefficient for a given x variable is interpreted as the amount of change in y corresponding to a one-unit change in that variable, when all other variables are held constant. Because b coefficients are also affected by the metric of the predictor variable, they are inappropriate for assessing the relative importance of multiple predictors within a regression. Thus, standardized regression coefficients (ß weights) are used as an alternative, leading to the following equation:

$$y = \beta_0 + \beta_1 x_1 + \beta_2 x_2 + \beta_3 x_3 + \cdots + \beta_k x_k$$

Where β_0 is the standardized intercept and $\beta_1 \ldots \beta_k$ are the standardized slope coefficients. Each standardized regression coefficient represents the amount of change in y corresponding to a one-standard deviation change in that variable, while statistically controlling for the other variables, which are also z-transformed (standardized to a mean of zero and a standard deviation of one). Typically, ß weights range from -1 to $+1$. The sign of the coefficient reveals the direction of the relationship between a given predictor variable and the dependent variable. When the standardized coefficient is close to zero, it reveals that the given variable contributes very slightly to the prediction. On the contrary, a standardized regression coefficient with a large absolute value indicates a greater contribution of the variable on the prediction of the dependent variable (Hatcher, 2013).

Multiple regressions used to model predictions rely on the assumptions that there is a linear relationship between each of the predictors and the dependent variable, that the effects of all independent variables on the dependent variable can be added together; that there is independence of the residual terms, that the residuals of prediction are approximately normally distributed and that there is homoscedasticity of residuals, i.e. that the variances along the regression line remain fairly similar for all values of x. Additionally, data should not show multicollinearity, a strong correlation among two or more predictors, because it leads to difficulties on understanding which of the independent variables contributes to the variance being explained and brings in some issues in the calculation of the model. Furthermore, the absence of outliers has to be checked for, as unusual observations that do not adhere to the model can introduce bias into the regression equation and thus reduce the accuracy of the results (Fahrmeir et al., 2013; Mertens et al., 2017). In the present study, these assumptions are tested using the SPSS software and in case of violation of any of the assumptions this will be explicated together with the related decisions.

Overall model fit statistics indicate whether the specified model fits the data. In other words, they reveal if the combination of independent variables and dependent variable matches the actual characteristics of the observed data. It contains two relevant measures. The first is the R^2 coefficient, which measures the amount of variance in the dependent variable that can be explained by the set of specified predictors. The second is an F statistic obtained by comparing the estimated model against a model in which all the coefficients are set to zero, i.e. against a model with no predictors. If the F statistic is significant, the estimated model is substantially better at predicting the dependent variable than a model based just on chance.

When a large number of predictor variables is available, a decision needs to be made as to which of these variables should be included in the regression model. Different approaches have been developed for model choice and variable selection matter, e. g. forward selection, backward elimination and stepwise selection. Fahrmeir et al. (2013) suggest that the different potential models can be compared using the global model choice criterion. To do this, relationships between the independent variables with the dependent variables are first analyzed independently. From these analyses, the most relevant predictors are identified. All possible models including these predictors are then calculated and compared using a global goodness-of-fit criterion, the AIC (Akaike Information Criterion). Finally, the model with the smallest AIC is selected.

3.3.2.7 Multinomial Logistic Regression Analysis

The linear multiple regression procedures described in the previous section are appropriate for predicting continuous dependent variables. However, when the variable to be predicted is categorical rather than continuous, a different method is needed. Multinomial logistic regression offers such an alternative, in which the aim is not determining the predicted value of the dependent variable, but the probability of being in a certain category of the dependent variable considering a set of predictors. In this study, multinomial logistic regression is used to model the relationship between the preferences for dealing with students' errors and several independent variables and also to model the relationship between the patterns of change in the preferences for dealing with students' errors and the independent variables.

Multinomial logistic regression seeks to estimate the probability of being in a particular category depending on the values of the independent variables, which can be categorical or numerical. Since probabilities are involved, the assignment to the categories is best represented by a logistic function. The model parameters for a logistic regression are estimated using maximum likelihood estimation (MLE), similarly as it is done in the item response theory already presented. A big difference is though, that in multinomial logistic regression more than one logistic regression equation is computed. Precisely, one equation less than the number of categories in the dependent variable because one of the categories is chosen as the reference category, against which the other two categories are compared. Then, a logistic regression equation is estimated for each comparison as follows (Hutcheson & Hutcheson, 2011):

$$log \frac{Pr(Y = j)}{Pr(Y = j')} = \beta_0 + \beta_1 x_1 + \beta_2 x_2 + \beta_3 x_3 + \cdots + \beta_k x_k$$

Where j is the category being compared and j' is the reference category. For each equation indices of model fit, logistic regression coefficients, adjusted odds ratios and related statistics are computed. First, to determine if there is a statistically significant relationship between the categorical dependent variable and the set of independent variables, a model χ^2 statistic is computed. If it is significant, the null hypothesis stating that the regression coefficients for all predictor variables are equal to zero can be rejected and it can be concluded that the model containing the constant and the set of predictor variables provides a significantly better fit to the data than the model with only the constant. Additionally, logistic regression coefficients for each predictor are estimated. These coefficients represent the change in the log odds associated to a one-unit change in the predictor variable when all other predictors are statistically held constant (Hatcher, 2013). Furthermore, the adjusted odds ratio is calculated, which "estimates the multiplicative change in the odds of membership in the targeted group for every one-unit increase in the predictor variable while statistically controlling for the other predictor variables" (Hatcher, 2013, p. 330).

As for multiple linear regression, when a number of potential predictors are available to be introduced in a multinomial logistic regression, decisions need to be made regarding to which independent variables should be included in the model. Here, the purposeful selection approach from Hosmer, Lemeshow and

Sturdivant (2013) was followed. It provides a set of criteria to find the most parsimonious model that will represent the data and minimize the estimated standard errors. It is not an automatized process conducted by a software itself but includes the researcher's specialized knowledge for reviewing and evaluating the model. In addition to Hosmer and colleagues' (2013) guidelines, information criteria were considered as an additional indicator for making relevant decisions.

Results

4

This chapter presents the results of the analyses carried out to investigate the research question and the hypotheses of the study. The first section exposes the results of the cross-sectional analyses. These were undertaken to explore the relevant characteristics of the diagnostic competence of the participating preservice primary school teachers and to examine the connections of those characteristics with other features of preservice teachers' background, such as professional knowledge, beliefs and practical experience. The second section presents the results of the longitudinal analyses conducted, which focused on the development of the diagnostic competence of preservice primary school teachers after participating in the university seminar sequence. In addition, this section analyzed the associations of changes in the development of the competence with the features of preservice teachers' background in order to identify aspects that can play an important role in the development of the competence within initial teacher education.

4.1 Cross-sectional Results

In this section, cross-sectional results about the diagnostic competence in error situations of preservice primary school teachers will be presented. In particular, two facets will be examined to investigate this competence, namely the competence to hypothesize about causes of errors and the preferences for dealing with student errors.

Each preservice teacher participating in the present study was assigned, both for the pre- and post-test, a person parameter estimating their skill level for the hypothesizing-about-causes competence. Also, every participant was assigned to a particular preference class for handling errors at each testing time. This allows examining the developmental level of the competence prior to and after participation in the university seminar sequence.

Additionally, relationships between both components of the diagnostic competence and of each of them with other background characteristics were investigated using the data obtained prior to the seminar sequence. In this way, the conceptualization of preservice teachers' diagnostic competence in error situations can be deepened and correlations with other variables can be suggested.

4.1.1 Hypothesizing-about-causes Competence

As described in chapter 3, the competence to hypothesize about causes of students' errors was measured with closed items. Using IRT analysis, person parameters for each participant were obtained and standardized to a mean of 50 and a standard deviation of 10. The range of these T-scores at the pre-test was 54.61, with a minimum of 14.42 and a maximum of 69.04.

The median (Mdn = 50.8899) was slightly higher than the mean, indicating that the distribution might be negatively skewed. This was confirmed by a negative and significant skewness statistic (Skewness = $-.731$, SE = .212, z = 3.45, p<.05). Also, the Kurtosis statistic showed some signs of distancing from the normal distribution (Kurtosis = 1.009, SE = .420, z = 2.40, p<.05), indicating a leptokurtic distribution, i.e. a distribution with a relatively high peak. Although these two characteristics can be observed in the histogram in Figure 4.1, an approximately normal distribution can also be identified.

Normality tests can also be used to assess normality. The null hypothesis in these tests is that the sample distribution is normal, so if the test is statistically significant, the distribution should not be assumed normal. Both the Shapiro-Wilk and the Kolmogorov-Smirnov tests suggest that the null-hypothesis should be rejected for this sample (p = .001 and p = .020, respectively). Graphical assessment can be used as a supplement to these tests. Examination of the boxplot in Figure 4.2 shows a symmetrical box with quartiles 1 and 3 approximately the same distance from the median and whiskers of a similar length. Extremes values are only observed at the lowest values. Similarly, the Q-Q plot in Figure 4.3 shows that most of the data points in the sample follow the normal distribution line, except for a few points at the bottom end. These few cases may be causing the normality tests significant results. For these reasons, normality of the data will be assumed but treated with caution in the following analyses.

Answers provided by participants after the seminar sequence also resulted in ability parameters for each participant. These post-test competence parameters

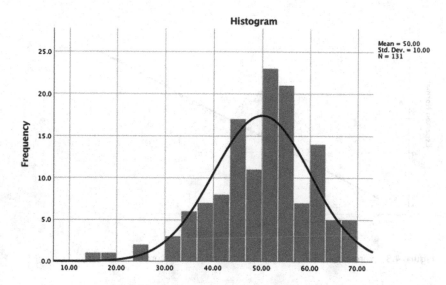

Figure 4.1 Histogram of the pre-test standardized scores

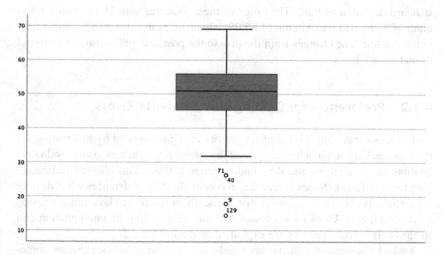

Figure 4.2 Boxplot of the pre-test standardized scores

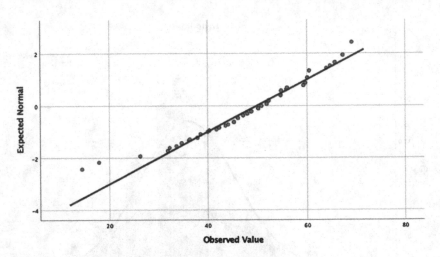

Figure 4.3 Normal Q-Q Plot of the pre-test standardized scores

were standardized using the first testing-time mean and standard deviation, as described in chapter 3. The mean at the second testing-time was 52.57, with a standard deviation of 9.89. The range of these T-scores was 51.44, with a minimum of 33.64 and a maximum of 85.08, which are both substantially higher than in the pre-test. The changes from the pre- to the post-test will be further analyzed in section 4.2.1.

4.1.2   Preferences for Dealing with Students' Errors

Preferences for dealing with students' errors were investigated by first coding the decision-making approaches suggested by preservice teachers using evaluative qualitative text analysis and then finding latent classes. This allowed distinguishing three different classes of teachers according to their preferences when dealing with students' errors: a constructivist class, an instructivist class and a missing evidence class. Each of these classes, the model selection, its interpretation and the answers' coding process are explained in section 3.3.2.2.

Table 4.1 summarizes the number and percentage of preservice teachers ordered into each of the classes in both testing-times. It can be seen that at the first test point, the majority of participants hold instructivist preferences, 22.9% revealed a constructivist orientation and over one quarter provided too little information about how they would handle the error. The percentage of instructivist-oriented preservice teachers decreased noticeably after the university seminar sequence and, conversely, the number of participants revealing constructivist preferences about dealing with errors increased. A closer examination of this change in error handling preferences can be found in section 4.2.2.

Table 4.1 Pre- and post-test class assignment according to dealing-with-errors preferences

	Pre-test		Post-test	
Preference class	Frequency	Percent	Frequency	Percent
Instructivist	67	51.1	50	38.2
Constructivist	30	22.9	41	31.3
Missing evidence	34	26.0	40	30.5
Total	131	100.0	131	100.0

4.1.3   Relationship between Both Competence Components

The relationship between the two components under investigation is of clear interest to have a complete depiction of the diagnostic competence in error situations. Table 4.2 shows how both aspects are related. It can be seen that the group showing constructivist preferences achieved the highest mean in the competence to hypothesize about causes of errors. Preservice teachers who exhibited instructivist preferences to dealing with errors reached a slightly lower mean and the lowest mean was the one from the missing evidence group.

Table 4.2 Descriptive statistics of the hypothesizing-about-causes competence for each class of dealing-with-errors preferences

Preference class	N	Minimum	Maximum	Mean	Std. Deviation	Std. Error
Instructivist	67	25.89	69.30	50.28	8.77	1.07
Constructivist	30	34.48	67.49	51.48	8.46	1.55
Missing evidence	34	13.90	67.49	48.14	13.10	2.25
Total	131	13.90	69.30	50.00	10.00	.87

In order to determine if the differences in the means of the three groups are statistically significant, a one-way Welch ANOVA was conducted. Examination of normal Q-Q Plots revealed two outliers on the missing evidence group and one on the instructivist group. Despite this, data were approximately normally distributed for each group, as evaluated by the Shapiro-Wilk test ($p > .05$). Because there was heterogeneity of variances, as assessed by Levene's test of homogeneity of variances ($p = .023$), the Welch statistic was used. The adjusted F ratio shows that differences between the three groups were not statistically significant, Welch's $F(2, 61.752) = .747$, $p = .478$. According to these results, it is not possible to establish a relationship between preservice teachers' preferences when dealing with students' errors and their competence to hypothesize causes for these errors.

4.1.4 Relationship between Diagnostic Competence and Preservice Teachers' Characteristics

To better understand diagnostic competence in error situations, its association with various background characteristics of preservice teachers was explored. For this purpose, the characteristics detailed in the research question and hypothesis chapter were selected, which include various features about the beliefs, knowledge, and teaching experience of future teachers. Each of these associations is examined below.

Relationship with Beliefs
Given the known relevance of teachers' beliefs for their practices, it is of interest to examine the connections between preservice teachers' diagnostic competence in error situations and their beliefs. Relations of each of both components of the diagnostic competence with participants' beliefs about the nature of mathematics and about teaching and learning of mathematics were investigated.

To test the association of the hypothesizing-about-causes competence with the beliefs, Pearson's correlations were run, in which pre-test person parameters of the competence and ipsative values of the beliefs were used. There was a statistically significant, moderate positive correlation between the hypothesizing competence and beliefs about the nature of mathematics as a process of inquiry, $r = .378$, $p = .000$. Similarly, a significant and moderate positive correlation with constructivist beliefs about the learning of mathematics was found, $r = .413$, $p = .000$. This means that the more constructivist participants' beliefs are, the higher scores they show in their competence to hypothesize about the causes of student errors. On the contrary, preservice teachers holding beliefs about the nature of mathematics as the application of a set of rules and procedures or about the learning of mathematics as a teacher-directed process exhibit lower personal parameters in the assessed hypothesizing competence.

Regarding the preferences for dealing with student errors, it is interesting to investigate their connection with preservice teachers' beliefs as constructivist and instructivist orientations are already included in the conformation and description of the preference classes. Conceptually, instructivist preferences for dealing with student errors are closely connected to beliefs about learning mathematics through following teacher direction as both focus on teachers' actions and rely on the assumptions that students should listen to teacher explanations and follow their instructions to learn mathematics. Similarly, they are connected to viewing mathematics as a set of rules and procedures that need to be transferred by teachers and learned by students in order to know the correct formula or procedure that would allow them to solve a problem. On the contrary, constructivist preferences for dealing with errors should be associated with beliefs about learning mathematics through active involvement, since both consider students as the main actors in their learning and support the idea that, for the learning process to be effective, students should be the ones who do the mathematics, engage intellectually in the tasks, work on their inquiries and develop their own strategies. Correspondingly, constructivist preferences should be associated with views of mathematics as a flexible process that permits individual trials and discoveries, as a process of inquiry and as a tool that allows solving real world problems.

To evaluate these connections, the data were summarized in Table 4.3 and analyses of variances were conducted. First, a one-way ANOVA was run to assess if the beliefs about the learning of mathematics varied among groups with different preferences for dealing with errors. Although the Shapiro-Wilk test suggests that this variable is significantly departed from a normal distribution in the constructivist group, examination of the histogram, box plot and normal Q-Q plot show an approximately normally distributed variable. Considering the robustness of the ANOVA procedure and that the assumption of homogeneity of variances was met, as confirmed by Levene's test ($p = .495$), the results were considered. They confirmed that constructivist beliefs about the learning of mathematics diverge significantly between participants with different preferences to dealing with student errors $F(2, 128) = 3.580, p = .031$, partial $\eta^2 = .053$.

Tukey HSD post-hoc analysis revealed significantly different means between the group with constructivist preferences and the missing evidence group ($p = .031$) and weakly significant differences between the constructivist and instructivist groups ($p = .079$). Differences between the instructivist and missing evidence groups were statistically not significant ($p = .723$).

In the same way, a one-way ANOVA was conducted to determine if the adherence to beliefs about the nature of mathematics as a process of inquiry was different for the three classes of preferences for dealing with errors. Examination of normal Q-Q plots revealed no outliers; data were normally distributed for each group, as assessed by the Shapiro-Wilk test ($p > .05$); and there was homogeneity of variances, as assessed by Levene's test of homogeneity of variances ($p = .282$). Constructivist beliefs about the nature of mathematics do differ significantly between participants with different decision-making preferences, $F(2, 128) = 7.875$, $p = .001$, partial $\eta^2 = .110$. Post-hoc test (Tukey HSD) indicates significantly different means in the beliefs between the groups with instructivist and constructivist preferences for dealing with student errors ($p = .000$) and between the constructivist and missing evidence groups ($p = .013$).

Table 4.3 Descriptive statistics of both beliefs aspects for each class of the dealing-with-errors preferences

Preference class	Beliefs about the nature of mathematics			Beliefs about learning mathematics		
	N	Mean	Std. Deviation	N	Mean	Std. Deviation
Instructivist	67	.33	.50	67	1.26	.52
Constructivist	30	.80	.63	30	1.55	.75
Missing evidence	34	.40	.59	34	1.16	.64
Total	131	.46	.58	131	1.30	.62

These results support the selection and interpretation of the preferences classes. They provide evidence in favor of the conceptual assumptions that constructivist preferences for dealing with student errors are related to constructivist beliefs about teaching and learning mathematics and about the nature of mathematics, in that for both beliefs aspects, the constructivist preferences group exhibited the highest parameters. The opposite holds too, the group with instructivist preferences for dealing with errors displays significantly lower parameters in the two measured beliefs areas.

Relationship with Professional Knowledge
Mathematical knowledge for teaching was assessed with a paper and pencil test. Individual ability parameters were calculated using IRT, as described in section 3.3.2.1. The relationship of this measurement of professional knowledge with each component of the diagnostic competence was investigated.

To assess the relationship of this metric variable with the, also numerical variable, hypothesizing-about-causes competence, Pearson's correlation was calculated. A statistically significant, moderate positive correlation was found between participants' mathematical knowledge for teaching and the competence to hypothesize about causes of students' errors, $r = .307$, $p = .000$. This result confirms that participants with a higher knowledge base also show a higher competence to think about possible causes for student errors and that those manifesting a weaker knowledge base also show a lower hypothesizing competence.

Table 4.4 Descriptive statistics of Mathematical Knowledge for Teaching for each class of the dealing-with-errors preferences

Preference class	N	Minimum	Maximum	Mean	Std. Deviation	Std. Error
Instructivist	67	27.12	66.32	50.36	8.50	1.04
Constructivist	30	27.12	77.58	53.32	13.72	2.50
Missing evidence	34	29.31	66.32	48.94	8.31	1.43
Total	131	27.12	77.58	50.67	9.93	.87

Table 4.4 displays the mathematical knowledge for teaching for each of the preference classes for dealing with student errors. It can be seen that the group with constructivist preferences exhibited the highest knowledge, followed by the instructivist-preferences group and the missing evidence group showed the weakest performance. To test if these differences are statistically significant, a one-way Welch ANOVA was conducted. There were no outliers, as assessed by visual inspection of a boxplot; data were normally distributed for each group, as evaluated by examination of normal Q-Q plots and Shapiro-Wilk test ($p > .05$); but the assumption of homogeneity of variances was not met, as revealed by Levene's test ($p = .006$). The adjusted F ratio indicates that the differences between the three dealing-with-errors-preference groups are not statistically significant, Welch's $F(2, 59.531) = 1.164$, $p = .319$. Therefore, there is no relationship between preservice teachers' professional knowledge and their preferences for dealing with student errors.

Relationship with University Entrance Test Score

Preservice teachers' university entrance test scores give a standardized measure of their academic abilities. The Chilean national mean varies slightly every year but is around 500 points with a standard deviation of 110. Similarly, the average scores of those who enroll in teaching education vary each year but are around 580, with a standard deviation of 71 (Arias & Villarroel, 2019). The mean of the participants in this study was 607.85, with a standard deviation of 47.35. In other words, the sample in the present study obtained higher scores than the population in the university entrance test.

A Pearson's correlation was run to assess the relationship between the university entrance results and the parameters of the hypothesizing-about-causes competence. There was a statistically significant, small positive correlation between both variables, $r = .276$, $p < .000$. Therefore, it can be stated that higher scores in the university entrance test are associated with a higher competence to make hypotheses about the causes of student errors, and vice versa.

Table 4.5 summarizes the mean score on the university entrance test for each of the dealing-with-errors preference groups.

Table 4.5 Descriptive statistics of the university entrance test scores for each class of the dealing-with-errors preferences

Preference class	N	Minimum	Maximum	Mean	Std. Deviation	Std. Error
Instructivist	67	486	715	606	47.2	5.8
Constructivist	30	540	730	625	52.3	9.5
Missing evidence	34	500	668	597	39.9	6.8
Total	131	486	730	608	47.4	4.1

The differences between the scores of the three groups were examined using a one-way ANOVA. It revealed that university entrance test scores were weakly statistically significantly different between the three preference groups, $F(2, 128) = 2.894$, $p = .059$, partial $\eta^2 = .043$. Tukey HSD post hoc analyses revealed a statistically significant difference between the scores of participants with constructivist preferences for dealing with student errors and those of the missing evidence group ($p = .051$). No other group differences were statistically significant.

Relationship with Study Progress

As it could be conjectured that with more completed semesters in a teacher education program, preservice teachers would have had more opportunities to learn and to develop professional competencies, the relationship between their study progress, measured by the number of passed semesters, and their diagnostic competence was examined.

First, a Pearson's correlation was run to test the relationship of the number of completed semesters at university and preservice teachers' competence to hypothesize about causes of student errors. There was a very weakly statistically significant positive correlation between both variables, $r = .121$, $p = .084$ (one-tailed).

Regarding the relationship of participants' study progress with their preferences for dealing with student errors, the data (summarized in Table 4.6) were significantly departed from a normal distribution in the instructivist preferences and the missing evidence groups, as determined by visual inspection of their histograms, normal Q-Q plots and box plots and by the Shapiro-Wilk's test ($p < .05$). There was also heterogeneity of variances, as stated by Levene's test of homogeneity of variances ($p = .017$). For these reasons, a non-parametric test, the Kruskal-Wallis test, was conducted. It suggests there is a significant difference between the groups, $X^2 (2) = 13.40$, $p = .001$. In order to compare—considering the robustness of the analysis of variances test—a one-way Welch ANOVA was conducted despite the violation of the assumptions described above. It also suggested significant differences between the groups $F(2, 64.266) = 10.447$, $p = .000$, est. $\omega^2 = .126$. Post-hoc Games-Howell test indicates significantly different means between the group with constructivist preferences and the missing evidence group ($p = .011$) and between the constructivist and instructivist preference groups ($p = .000$).

Table 4.6 Descriptive statistics of the number of passed semesters for each class of the dealing-with-errors preferences

Preference class	N	Mean	Std. Deviation	Std. Error
Instructivist	67	5.15	2.245	.274
Constructivist	30	7.20	2.007	.366
Missing evidence	34	5.32	2.962	.508
Total	131	5.66	2.529	.221

Relationship with Mathematics or Mathematics Education Courses
To explore deeper the relationship between opportunities to learn in teacher education and the development of the diagnostic competence, the association between

the number of courses of mathematics, mathematics education or didactics of mathematics passed by preservice teachers and each component of the diagnostic competence was investigated.

The association between the number of passed mathematics or mathematics education courses and preservice teachers' competence to hypothesize about causes of student errors was evaluated using a Pearson's correlation analysis. There was a statistically significant, small positive correlation between both variables, r = .144, p = .050 (one-tailed). This means that participants who had completed more mathematics or mathematics education courses tend to show a higher ability to make hypotheses about possible causes of students' errors.

The relationship between the number of mathematics or mathematics education courses participants have completed within their university education and their preferences for dealing with student errors was also investigated. Visual inspection of histograms, normal Q-Q plots and box plots and a Shapiro-Wilk's test (p <.05) of the data summarized in Table 4.7 showed that the dependent variable is departed from a normal distribution in the instructivist and missing evidence preference groups. This led to conducting the non-parametric Kruskal-Wallis test, which revealed a significant difference between the groups X^2 (2) = 22.43, p = .000. Additionally, considering the robustness of the test and that the assumption of homogeneity of variances was met (Levene's F(2,128) = 1.776, p = .173), a one-way ANOVA was conducted to compare the number of mathematics or mathematics education courses participants with different decision-making preferences have completed. This test also showed a significant difference in the number of courses participants with different decision-making preferences have completed, F(2, 128) = 13.412, p = .000, partial η^2 = .173. The Tukey HSD post hoc test indicates significantly different means between the constructivist and missing evidence preference groups (p = .000) and between the constructivist and instructivist preferences groups (p = .000).

Table 4.7 Descriptive statistics of the number of passed mathematics and mathematics education courses for each class of the dealing-with-errors preferences

Preference class	N	Mean	Std. Deviation	Std. Error
Instructivist	67	3.34	1.647	.201
Constructivist	30	5.43	2.208	.403
Missing evidence	34	3.26	2.300	.395
Total	131	3.80	2.146	.188

Relationship with School Practicum
Another aspect of opportunities to learn is time student teachers spend in field placements. Typically, in this kind of practical experience in Chilean universities, preservice teachers would be involved in school once a week and would start by having an observing role in the first year and move towards a more active role in the second and third years. In the fourth year, an internship component is added, where student teachers would be at school more than three times a week. For this study's purposes, any of these field placements lasting for a whole semester, despite weekly frequency, was counted as one practicum.

The relationship of the number of these one-semester practicums with the competence to hypothesize about causes of student errors was analyzed conducting a Pearson's correlation. It showed that there is a statistically significant, small positive correlation between both variables, $r = .164$, $p = .031$ (one-tailed). In other words, preservice teachers who have more school practicum experiences usually exhibit also a higher competence to hypothesize about causes of student errors.

Table 4.8 shows the correlation between the number of school practicum completed by preservice teachers and their preferences for dealing with student errors. Preliminary analyses showed that the dependent variable was not normally distributed for each of the preference groups. Visual inspection of the histograms, normal Q-Q plots and box plots and Shapiro-Wilk's test results ($p < .05$) indicated that the number-of-school-practicum variable was departed from a normal distribution in the instructivist preferences and missing evidence groups. There was also heterogeneity of variances, as stated by Levene's test ($p = .000$). Considering this situation, the Kruskal-Wallis non-parametric test was conducted. It revealed there is not a significant difference between the groups $X^2 (2) = 18.50$, $p = .397$. Although some of the assumptions for a one-way ANOVA were not met, considering the robustness of the test and with the aim of comparing results, the analysis of variances was conducted. It also showed no statistically significant relationship between the number of school practicum preservice teachers with different dealing-with-errors preferences have experienced, $F(2, 128) = 1.563$, $p = .213$. Thus, within these data it is not possible to identify an association between preservice teachers' practicum experiences and their preferences for dealing with errors.

Relationship with Teaching Experience in Primary Classrooms
To explore deeper the relationship between practical experiences and the development of the diagnostic competence, preservice teachers were asked about the frequency with which they had taught single lessons in primary classrooms. Participants' answers were summarized into two categories: one with participants with no

Table 4.8 Descriptive statistics of the number of school practicum for each class of the dealing-with-errors preferences

Preference class	N	Mean	Std. Deviation	Std. Error
Instructivist	67	3.69	2.500	.305
Constructivist	30	4.30	1.725	.315
Missing evidence	34	3.24	2.720	.466
Total	131	3.71	2.419	.211

experience or with a single lesson experience, and a second category with participants ranging from some experience to very frequent teaching. An independent-samples t-test was run to determine if there were differences in the development of the hypothesizing competence between the groups with and without experience teaching in primary classrooms. Competence parameters for each category were normally distributed, as assessed by Shapiro-Wilk's test ($p > .05$) and there was homogeneity of variances, as assessed by Levene's test ($p = .116$). As shown in Table 4.9, preservice teachers with teaching experience in primary classrooms exhibited a statistically significantly higher level ($M = 45.95, SD = 11.47$) of the hypothesizing about causes of student errors competence than participants without experience ($M = 51.65, SD = 8.88$), $t(129) = -3.054, p = .003$. According to Cohen's d, there is a medium size effect, $d = .555$ (Table 4.9).

Table 4.9 Descriptive statistics of the competence to hypothesize about causes of student errors for participants with and without teaching experience in primary classrooms

Teaching experience in primary classrooms	N	Mean	Std. Deviation	Std. Error
Without teaching experience	38	45.95	11.47	1.86
With teaching experience	93	51.65	8.88	.92

Table 4.10 shows the relationship between preservice teachers' experience in primary classrooms and their preferences for dealing with student errors.

The data show that the instructivist and missing evidence classes have similar teaching experience distributions. However, the analysis of this relationship also yielded a non-significant result, $X^2 (2, N = 131) = 1.544, p = .462$. Therefore, it is not possible to establish a relationship between the experiences preservice teachers have had teaching in primary classrooms and their preferences for dealing with student mathematical errors.

Table 4.10 Relationship between teaching experience in primary classrooms and preference for dealing with student errors

Preference class	Without teaching experience in primary classrooms	With teaching experience in primary classrooms
Instructivist	21	46
Constructivist	6	24
Missing evidence	11	23
Total	38	93

Relationship with Experience Teaching Mathematics in Primary Classrooms
It can be hypothesized that practical experiences of teaching specifically mathematics to primary school students might have an influence on the development of preservice teachers' diagnostic competence. To investigate the relationship between these two variables, preservice teachers were asked about the frequency with which they had taught mathematics lessons in primary classrooms. Their answers were summarized again into the two categories described in the previous section, namely one category for participants with no experience or with a single lesson experience, and a second category with participants ranging from some experience to very frequent teaching. Table 4.11 exhibits the relationship between preservice teachers' experiences teaching mathematics to primary school students and their competence to hypothesize about causes for student errors.

Table 4.11 Relationship between experience teaching mathematics in primary classrooms and preference for dealing with student errors

Experience teaching mathematics in primary classrooms	N	Mean	Std. Deviation	Std. Error
Without teaching experience	49	47.45	11.13	1.59
With teaching experience	82	51.53	8.99	.99

Although the Shapiro-Wilk test suggests that the competence parameters are departed from a normal distribution in the group without experience ($p = .028$), examination of histograms, box plots and normal Q-Q plots show an approximately normally distributed variable. Considering the robustness of t-tests and the fact that the assumption of homogeneity of variances was met, as confirmed by Levene's test ($p = .199$), the results were considered. The independent-samples t-test revealed a significant difference between the means of these two groups,

$t(129) = -2.297$, $p = .023$, with a medium size effect according to Cohen's d $= .517$. In other words, preservice teachers who have had experiences teaching mathematics more than once showed a statistically significant higher level of the competence of hypothesizing about causes of student errors ($M = 51.52$, $SD = 8.99$) than those who have not had the opportunity to teach mathematics ($M = 47.45$, $SD = 11.13$).

The relationship between preservice teachers' preferences for dealing with student errors and their experiences teaching mathematics in primary grades is summarized in Table 4.12.

Table 4.12 Relationship between experience teaching mathematics in primary classrooms and preference for dealing with student errors

Preference class	Without teaching experience in primary classrooms	With teaching experience in primary classrooms
Instructivist	27	40
Constructivist	8	22
Missing evidence	14	20
Total	49	82

A chi-square test of independence was conducted between preservice teachers' preferences for dealing with student errors as one variable and the two categories of without-mathematics-teaching-experience and with-mathematics-teaching-experience as the second nominal variable. There was not a statistically significant association between them, X^2 ($2, N = 131$) $= 1.924$, $p = .382$. In other words, with the available data, it is not possible to identify a relationship between preservice teachers' experiences teaching mathematics and their preferences for dealing with student errors.

Relationship with Private Tutoring Experience
The relationship between extracurricular opportunities to learn with the development of the diagnostic competence was also explored. Therefore, participants were asked about their experiences doing private tutoring. They provided information about each of the school grades for which they had given private lessons. With this information, it could be distinguished between participants with private tutoring experience for any grade level and those who had not done private lessons for students at any school grade.

Preservice teachers' competence to hypothesize about causes of student errors for groups with and without private tutoring experience is displayed in Table 4.13. The Shapiro-Wilk test suggests that the competence variable is significantly departed from a normal distribution; however, examination of the histograms, box plots and normal Q-Q plots indicates an approximately normally distributed variable. Since t-tests are robust and the assumption of homogeneity of variances was met, as confirmed by Levene's test ($p = .838$), the test was conducted. Results indicate that there is not a significant difference in the ability parameters from participants with (M $= 50.73$, SD $= 10.16$) and without (M $= 48.69$, SD $= 9.68$) private lesson experience; $t(129) = -1.121$, $p = .264$.

Table 4.13 Descriptive statistics of the competence to hypothesize about causes of student errors for participants with and without experience giving private lessons

Experience	N	Mean	Std. Deviation	Std. Error
Without private lessons experience	47	48.69	9.68	1.41
With private lessons experience	84	50.73	10.16	1.11
Total	131	50.00	10.00	.87

Interestingly, when preservice teachers are categorized according to their experience giving private lessons for primary grades students, results indicate there are significant differences. Table 4.14 displays this relationship. Although the Shapiro-Wilk test indicates that the data were departed from a normal distribution, visual examination of the histograms, box plots and normal Q-Q plots reveals an approximately normally distributed variable. As there is homogeneity of variances (Levene's $F(1,129) = .714$, p $= .400$) and t-tests are considered robust against some deviations from normality, the test was conducted. Results indicate preservice teachers with experience tutoring primary school students have a significantly higher competence to hypothesize about causes of student errors than those who have either no private tutoring experience or have tutored students of other school grades, t(129) $= 1.630$, p $= .053$ (one-tailed), Cohen's d $= 284$. In spite of the small effect of the relationship, it is worthwhile to consider the results as they point out to an association of private tutoring experiences with a higher level of the hypothesizing competence.

The relationship between preservice teachers' preferences for dealing with student errors and their experience giving private lessons was also investigated. Table 4.15 displays the association of the preferences with the experience giving private lessons to students from any grade level. The conducted chi-square test of

Table 4.14 Descriptive statistics of the competence to hypothesize about causes of student errors for participants with and without experience giving private lessons to primary school students

Experience	N	Mean	Std. Deviation	Std. Error
Without private lessons experience to primary students	66	48.60	9.09	1.12
With private lessons experience to primary students	65	51.43	10.73	1.33
Total	131	50.00	10.00	.87

independence indicated there was no a statistically significant association between these variables, X^2 (2, $N = 131$) = 4.272, $p = .118$.

Table 4.15 Relationship between experience giving private lessons and preference for dealing with student errors

Preference class	Without private lesson experience	With private lesson experience	Total
Instructivist	27	40	67
Constructivist	6	24	30
Missing evidence	14	20	34
Total	47	84	131

Similarly, the association between preservice teachers' preferences for dealing with student errors and their experience tutoring students from primary grades is displayed in Table 4.16. The connection between these two variables was also statistically not significant, X^2 (2, $N = 131$) = 4.023, $p = .134$. Therefore, it is not possible to determine a connection between preservice teachers' private tutoring experiences and the ways they prefer to handle student errors.

4.1.5 Regression Analyses

To better understand the relationship between diagnostic competence and the independent variables, regression analyses were conducted. A multiple regression analysis was used to study the association of the competence to hypothesize about causes of student errors with the predictor variables. On the other hand,

Table 4.16 Relationship between experience giving private lessons to primary students and preference for dealing with student errors

Preference class	Without private lesson experience to primary students	With private lesson experience to primary students	Total
Instructivist	34	33	67
Constructivist	11	19	30
Missing evidence	21	13	34
Total	66	65	131

a multinomial regression analysis was used to investigate the association of the classes of preferences for dealing with errors with the independent variables.

First, pre-test data were analyzed using multiple regression with simultaneous entry of the predictor variables. The competence parameter for hypothesizing about causes of student errors was the criterion variable in this analysis. Independent variables that showed a significant relationship with the competence in the explorative analyses were included as predictor variables. As there was multicollinearity between the university entrance test scores and the mathematical knowledge for teaching test scores, only the latter was included in the analysis as it provides more specific information in relation to a professional competence for mathematics teaching. Multicollinearity was also found between the number of school practicum and teaching experience in primary classrooms and between these two variables and experience teaching mathematics in primary classrooms. Again, the latter was selected because it provided more specific information. The other six independent variables were entered into the model, namely beliefs about the nature of mathematics, beliefs about the learning of mathematics, the number of mathematics or mathematics education university courses, mathematical knowledge for teaching, experience teaching mathematics in primary classrooms, and primary grades private lessons experience. The multiple regression model significantly predicted the competence to hypothesize about causes of student errors, $F(6, 124) = 5.889$, p $= .000$, adj. $R^2 = .184$. However, the regression on the different independent variables yielded only small effect sizes and only the beliefs about learning of mathematics variable added statistically significantly to the prediction ($\beta = .26$, p$< .05$). Regression coefficients and standard errors can be found in Table 4.17. These results mean that the more constructivist-oriented beliefs about how mathematics learning, the higher a preservice teacher scores on the competence to hypothesize about causes of student errors.

Table 4.17 Summary of Multiple Regression Analysis with all variables

	B (Unstandardized Coefficients)	Std. Error of B	β (Standardized Coefficients)	Significance
Constant	36.793	4.396		.000
Beliefs—nature of mathematics	3.143	1.803	.183	.084
Beliefs—learning of mathematics	4.126	1.651	.256	.014
Mathematical knowledge for teaching	.141	.097	.140	.148
Courses of mathematics or mathematics education	−.574	.483	−.123	.237
Experience teaching mathematics	1.042	1.975	.051	.599
Private lessons experience—primary	1.654	1.792	.083	.358

To find a more parsimonious model, following the suggestions of Fahrmeir et al. (2013), fourteen different regression models were examined (see Table 4.18), which do not differ in the inclusion of the variable beliefs about the learning of mathematics. Using the AIC as criteria, model number eight was selected. This model significantly predicts the competence under study, $F(3, 127) = 11.320$, p $= .000$, and it was also the one explaining the most variance (19.2%).

Table 4.19 shows the estimated coefficients and standard errors for the selected model. It reveals that the model contains only one significant regression coefficient that differs from zero at the 5% level: the constructivist beliefs about learning mathematics ($\beta = .26$, p<.05). In addition, two non-significant regression coefficients of small effect size explained some of the variance of the competence to hypothesize about causes of student errors. This means that, if both constructivist beliefs variables are controlled for, preservice teachers with higher mathematical knowledge for teaching also perform better in the hypothesizing about causes of errors test ($\beta = .14$, $p = .128$). Also, the more strongly constructivist beliefs about the nature of mathematics of preservice teachers are, the better their results in the hypothesizing test ($\beta = .17$, $p = .096$). Although these coefficients were not statistically significant at the 5% level, they make some contributions that also point in the expected direction.

Table 4.18 Potential models of a final regression analysis for the competence to hypothesize about causes of student errors

Model	Beliefs—Nature of mathematics	Beliefs—Learning of mathematics	Mathematical knowledge for teaching	Courses of mathematics or mathematics education	Experience teaching mathematics	Private lessons experience—primary	AIC	R^2
1	x	x	x	x	x	x	583.431	.184
2	x	x	x	x	x		582.328	.185
3	x	x	x	x			580.639	.190
4	x	x	x		x	x	582.914	.181
5	x	x	x			x	580.947	.188
6	x	x	x	x		x	581.725	.189
7	x	x	x		x		581.162	.186
8	x	x	x				579.231	.192
9	x	x					579.631	.184
10		x	x				580.102	.181
11	x	x			x		580.978	.181
12	x	x		x			581.496	.178
13	x	x				x	581.229	.180
14	x	x					581.767	.164

Table 4.19 Summary of Multiple Regression Analysis—parsimonious model

	B (Unstandardized Coefficients)	Std. Error of B	β (Standardized Coefficients)	Significance
Constant	36.436	4.287		.000
Beliefs—nature of mathematics	2.914	1.737	.170	.096
Beliefs—learning of mathematics	4.115	1.639	.255	.013
Mathematical knowledge for teaching	.136	.089	.135	.128

A second analysis was conducted to model the relationship between the preferences for dealing with student errors and a set of independent variables. Because the preferences classes are a nominal variable with three categories, a multinomial logistic regression analysis was chosen. For the selection of independent variables, the purposeful selection method from Hosmer et al. (2013) was followed, considering the AIC as an additional indicator, as it was done in the multiple regression analysis. First, all independent variables identified as significantly correlating with dealing with errors preferences during the explorative phase were considered and after evaluating for multicollinearity, four variables were entered into the model: beliefs about the nature of mathematics, beliefs about the learning of mathematics, the number of mathematics or mathematics education university courses and university entrance test scores. This initial model significantly improved the null model to predict preservice teachers' preferences for dealing with student errors, X^2 (8, $N = 131$) = 30.976, $p = .000$. As Table 4.20 reveals, statistically significant contributions were made only by the number of mathematics or mathematics education university courses. Examination of the parameter estimates at this stage revealed a significant effect of this variable on showing constructivist preferences for dealing with errors rather than instructivist preferences ($\beta = .637$, $p = .002$) or belonging to the missing evidence group ($\beta = .653$, $p = .008$). Although non-significant, it is worth noting that constructivist beliefs about the nature of mathematics tend to increase the likelihood of showing constructivist rather than instructivist preferences for dealing with student errors ($\beta = .323$, $p = .064$).

Table 4.20 Likelihood ratio tests for initial multinomial regression model

	AIC of reduced model	−2 log likelihood of reduced model	Chi-square	df	Significance
Intercept	256.657	240.657	1.624	2	.444
Beliefs—nature of mathematics	259.539	243.539	4.506	2	.105
Beliefs—learning of mathematics	256.380	240.380	1.347	2	.510
University entrance test score	255.850	239.850	.817	2	.665
Courses of mathematics or mathematics education	267.434	251.434	12.401	2	.002

After examining various models and checking the relationships among the variables as suggested by Hosmer and colleagues (2013), a final model was obtained. This final model included only two predictors, the constructivist beliefs about the nature of mathematics and the number of university courses of mathematics or mathematics education and it outperformed the null model, X^2 (4, $N = 131$) = 28.411, $p = .000$. As Table 4.21 reveals, only the variable courses of mathematics or mathematics education contributed significantly to the model at the 5% level.

Table 4.21 Model fitting information for final multinomial regression model

	AIC of reduced model	−2 log likelihood of reduced model	Chi-square	df	Significance
Intercept	241.482	233.482	35.226	2	.000
Beliefs—nature of mathematics	211.517	203.517	5.261	2	.072
Courses of mathematics or mathematics education	219.816	211.816	13.560	2	.001

However, when looking closer at the parameter estimates in Table 4.22, it can be seen that both predictors had significant parameters for comparing the constructivist group with the instructivist group. Estimates suggest that individuals with stronger constructivist beliefs about the nature of mathematics as an inquiry process are significantly less likely to belong to the instructivist than to the constructivist group. Similarly, the more university courses on mathematics and mathematics education preservice teachers have completed, the less likely they are to hold instructivist preferences for dealing with student errors rather than constructivist preferences. On the other side, only one predictor had significant parameters for comparing the constructivist group with the missing evidence group. The model suggests that participants who have completed more courses on mathematics and mathematics education are less likely to be in the missing evidence group than showing constructivist preferences. Constructivist beliefs had no significant effect on the likelihood of having constructivist preferences rather than showing poor evidence.

Table 4.22 Parameter estimates of the final multinomial regression model

	Instructivist preferences (N = 67)				Missing evidence (N = 34)			
	B	Std. Error of B	Sig.	β	B	Std. Error of B	Sig.	β
Intercept	3.217	.684	.000		2.586	.728	.000	
Beliefs—nature of mathematics	−1.053	.476	.027	.349	−.757	.526	.150	.469
Courses of mathematics or mathematics education	−.429	.140	.002	.651	−.476	.158	.003	.621

4.1.6   Discussion of Cross-sectional Results

The results presented so far allowed exploring the relationship between preservice primary school teachers' diagnostic competence and the independent variables and, therefore, to investigate the second hypothesis of this study. However, because multiple significance tests were conducted with the same data set to prove this hypothesis, the problem of multiple testing needs to be taken into account, as described in section 3.3.2.4. To apply the Holm-Bonferroni correction, the two facets of the diagnostic competence were considered separately, so the correction was applied once for the set of tests exploring the relationship of the hypothesizing about causes competence with other variables and another time to the set of tests investigating the association of the preferences for dealing with student errors and the independent variables. After the correction, most of the relationships remained statistically significant, especially those that also contributed significantly to the regression analyses. Hereafter, the corrected significance levels will be considered for further discussion. Despite the correction for multiple testing, results must be interpreted with caution as this is an exploratory study with a small and non-probabilistic convenience sample. Therefore, it is unlikely that the sample is representative of the population being studied, and generalizations cannot be made. Results should be interpreted as hints that should be examined in further studies.

The first relationships under examination were between preservice teachers' constructivist beliefs and their diagnostic competence in error situations. It was expected that constructivist beliefs about the nature of mathematics, i.e., viewing mathematics as an inquiry process, would be related to more flexibility and a higher competence in the search for causes for student errors, in contrast to a view of mathematics as a set of fixed rules and procedures, that would be associated to a narrower ability to find causes for student errors. Additionally, considering that under a constructivist perspective of the learning of mathematics, errors become more important and play a role in promoting students learning, it was also conjectured that participants with such beliefs would show a higher competence to hypothesize about causes of student errors. Both assumptions were confirmed by the data. Results showed a statistically significant difference, with a medium effect between participants' parameters of the competence to hypothesize about causes of student errors and the ipsative values of the constructivist beliefs both about the nature of mathematics and about the learning of mathematics. These results suggest that stronger constructivist beliefs are associated with higher levels of the competence to hypothesize about causes of student errors, which was also found in Heinrichs' study (2015) with preservice secondary mathematics teachers.

Regarding the relationship between preservice teachers' preferences for dealing with student errors and their beliefs, statistically significant results were also found. However, after correcting for multiple testing, only the beliefs about the nature of mathematics as an inquiry process show a significant association. This beliefs variable is also included in the final multinomial regression model, suggesting that preservice teachers with beliefs about the nature of mathematics as a process of inquiry are significantly more likely to belong to the group with constructivist preferences for dealing with errors and those with beliefs about the nature of mathematics as a set of rules and procedures are more likely to show instructivist preferences for dealing with student errors. This result confirms the conformation of the preference groups obtained with the latent class analysis, as the description of the instructivist and constructivist preferences groups are closely related to these two views of the nature of mathematics.

The second explored relationship was between preservice teachers' professional knowledge, namely their mathematical knowledge for teaching, and their diagnostic competence. It was hypothesized that a stronger knowledge base about mathematics teaching would be associated with a stronger ability to interpret student thinking and to look for different sources for their mathematical errors. In fact, a statistically significant relationship was found between participants' parameters in the mathematical knowledge for teaching test and their competence to hypothesize about causes of student errors. This moderate effect remained significant after the multiple testing correction and was, although not-significant, also relevant in the multiple regression analysis. This suggests that there is an association between preservice primary teachers' professional knowledge and their ability to think about possible causes for students' mathematical errors.

On the other hand, there was no significant relationship between participants' mathematical knowledge for teaching and their preferences for dealing with students' errors. This result could be explained by the assumption that stronger knowledge about teaching mathematics may have led preservice teachers to show a tendency towards a certain way of teaching mathematics. However, results did not confirm this hypothesis, suggesting that stronger mathematical knowledge for teaching could be associated with different preferences for dealing with errors in the primary school classroom.

The third hypothesis was aimed at investigating the connection between overall academic abilities and knowledge and the development of the diagnostic competence. Preservice teachers' scores in the university entrance test were taken as a standardized measure of their academic performance. A rather small relationship, but still statistically significant after the Holm-Bonferroni correction, was found between the competence to hypothesize about causes of students' errors and their academic abilities. This relationship was weaker than the one with the professional knowledge measure. Although no causal relationship can be established, this may be an indicator of the greater relevance of specialized professional knowledge over general academic abilities for the development of the hypothesizing feature of the diagnostic competence.

On the other hand, the relationship between academic performance and the preferences for dealing with errors was not statistically significant after the multiple testing adjustment. Therefore, with the available data, no significant disparities could be recognized in the academic performance of preservice teachers with different preferences for dealing with student errors.

The fourth hypothesis being tested was about the relationship between preservice teachers' progress in their university studies and the development of their diagnostic competence. This hypothesis is based on the assumption that preservice teachers in a higher semester have already had more learning opportunities to develop their diagnostic competence. Surprisingly, after the multiple testing correction, the association between the number of completed semesters and the hypothesizing-about-causes-of-errors competence was not significant, so no relation can be established between preservice teachers' progress in their studies and their ability to hypothesize about causes of students' errors. However, significant relationships were found with their specialized knowledge and their teaching experiences, so it is possible that the number of semesters may not have given enough information about the learning opportunities that allow them to specifically develop the skills needed to make hypotheses about the causes of student errors.

On the contrary, a Holm-Bonferroni corrected significant association was found between preservice teachers' progress in their studies and their preferences for dealing with student errors. Preservice teachers with constructivist preferences for dealing with student errors had completed significantly more semesters than those with instructivist preferences and those in the missing evidence group. This result was expected because preservice teachers in a higher semester have had more opportunities to learn about constructivist theory, not only in mathematics education courses but also in other disciplines and in psychology-related courses, so they may have internalized constructivist principles and they may have been capable of applying them into mathematics teaching and learning situations.

The fifth hypothesis refers to the relationship between the number of mathematics or mathematics university courses completed by preservice teachers and the development of their diagnostic competence. As by the fourth hypothesis, this was also based on the assumption that students who have completed more courses in the area of mathematics education have had more learning opportunities to develop their diagnostic competence. Although the relationship with the competence to hypothesize about causes of student errors was slightly significant with a small size effect, no significant association can be found after the multiple testing correction. It is possible to conjecture that this is due to the fact that knowledge about students errors and their analysis is not often included in Chilean university courses syllabus, so preservice teachers might have learned how to teach certain topics but not necessarily have had the opportunities to analyze possible causes of student errors in their mathematics education courses.

Conversely, the number of mathematics or mathematics education courses was closely associated with the preferences for dealing with student errors even after the Holm-Bonferroni adjustment. This variable also played a relevant role in the multinomial regression analysis. Preservice teachers with constructivist preferences for handling student errors reported they had completed a higher number of mathematics or mathematics education courses than those with instructivist preferences and those in the missing evidence group. It can be hypothesized that this might be due to the fact that in those courses, they have learned how to teach mathematics under a constructivist perspective, so the more courses they have completed, the more internalized they have constructivist principles of teaching and learning that they apply when suggesting ways to handle students' errors.

The sixth hypothesis states that there is an association between the number of school practicum preservice primary school teachers have done and the development of their diagnostic competence. It was assumed that more practical experiences would be connected to more opportunities to understand students' thinking and develop their diagnostic competence. Some studies highlight the relevance of teaching experience in the development of diagnostic abilities (Cooper, 2009). In the present study, after the multiple testing corrections, neither the competence to hypothesize about causes of student errors nor the preferences for dealing with student errors were significantly correlated with the number of school practicum preservice teachers have completed.

The seventh hypothesis aimed at exploring closer the relationship between practical experiences and the development of the diagnostic competence. In particular, it studied the association of this competence with preservice teachers' experiences teaching any subject in primary classrooms. As many school practicum are mainly composed of observing activities and preservice teachers have few opportunities to actually teach, it was of interest to investigate the relationship with their actual experiences teaching primary school children. After the multiple testing adjustment, significant differences in the competence to hypothesize about student errors were found between preservice teachers with and without teaching experiences in primary classrooms. On the contrary, no significant differences were found on the preferences for dealing with errors of preservice teachers with and without experience teaching in primary school classrooms. This may be due to the diversity of schools in which preservice teachers' practicums have taken place. Some schools and some in-service teachers may be working under a constructivist paradigm and others may be doing it under an instructivist one, so preservice teachers experience a variety of perspectives in school settings.

The next hypothesis is similar to the former one and states that there is a relationship between the development of the diagnostic competence and preservice teachers' experiences teaching specifically mathematics in primary school classrooms. One reason for this supposition was that frequent experiences teaching mathematics would lead to a higher mathematical pedagogical content knowledge due to the preparation process that teaching in the context of school practicum involves. It was also assumed that more frequent teaching would lead to a higher exposition to students' errors, their identification and interpretation, and therefore to more opportunities to develop the diagnostic competence. Although the association of the experiences teaching mathematics in primary classrooms with the hypothesizing about causes of student errors competence could not be confirmed with the sample in this study after the multiple testing correction, a significant and medium size effect relationship was found before the adjustment. Therefore, it would be of interest to further investigate this association in other studies.

In the same way as for the teaching experience in primary school classrooms, no significant association was found between preservice teachers' experiences teaching mathematics and their preferences for dealing with student errors. Again, the nature of their experiences in schools may not have influenced their teaching preferences in a particular direction.

The ninth hypothesis is related to the connection between preservice teachers' private tutoring experiences and the development of their diagnostic competence. As the previous hypotheses, it was based on the assumptions that practical experiences are relevant for the development of the diagnostic competence and that teaching experience would increase the frequency with which students' errors are faced and thus the opportunities to develop this competence. Also, Heinrichs (2015) found a significant relation between private tutoring and the competence to hypothesize about causes of student errors in her study with preservice secondary teachers. Although a similar relation was initially found in the present study, after the Holm-Bonferroni correction, the association between preservice primary school teachers' competence to hypothesize about causes of student errors and their experience giving private tutoring to students of any grade level or specifically to primary school students could not be confirmed. However, it is worth highlighting that the connection was stronger when preservice teachers were categorized according to their experiences tutoring specifically students who attended primary grade levels. Regarding the preferences for dealing with student errors, as in Heinrichs (2015) study, no statistically significant relation was found with the existence or absence of private tutoring experience.

To sum up briefly, after the multiple testing correction, the competence to hypothesize about causes of student errors was significantly associated to preservice primary school teachers' beliefs about the nature of mathematics and about the learning of mathematics, to their mathematical knowledge for teaching, to their university entrance test score and to their teaching experience in primary schools. The preferences for dealing with students' errors, after the multiple testing adjustment, were statistically significantly related to preservice primary teachers' beliefs about the nature of mathematics, to the number of mathematics or mathematics education courses and the progress they have made in their undergraduate studies.

4.2   Longitudinal Results

In this section, the changes in the development of the diagnostic competence of preservice teachers after their participation in the university seminar sequence are analyzed in detail. For this, the changes in the facet of the hypothesizing-about-causes-of-errors competence and in the facet of the preferences for dealing with student errors will be examined, which refers to hypothesis number 1 of the present study. In addition, considering that the effect of the university seminar sequence varied widely among participants, the changes in each of the facets will be related to several background characteristics of the preservice teachers. In this way, hypothesis 3 about the factors that can exert an influence on the changes in the development of the diagnostic competence will be examined.

4.2.1   Changes in the Hypothesizing-about-causes Competence

The changes in the development of the competence to hypothesize about causes for student errors experienced by preservice teachers in the frame of the university seminar sequence were examined using the person parameters obtained using IRT-scaling. As explained in chapter 3, person parameters from the first testing time were standardized to a mean of 50 and a standard deviation of 10. Post-intervention parameters were linearly transformed using the mean and standard deviation from the first testing-time.

A paired-samples t-test was used to determine whether there was a statistically significant change in the mean parameters of preservice teachers' hypothesizing-about-causes competence before and after the university seminar sequence. As a significant result can be noted that participants obtained significantly better results after the seminar sequence ($M = 52.6$, $SD = 9.89$) as opposed to before the seminar sequence ($M = 50$, $SD = 10$), a statistically significant mean increase of a small size effect, $t(130) = -2{,}649$, $p = .009$, $d = .231$.

4.2.2   Changes in the Preferences for Dealing with Students' Errors

Preservice teachers' preferences for dealing with student errors were categorized using latent class analysis, as explained in section 3.3.2.2, into three different classes: constructivist preferences, instructivist preferences and missing evidence. As stated in section 4.1.2. the vast majority of preservice teachers showed instructivist preferences prior to the seminar sequence (51.1%) and this proportion decreased considerably after the university seminar sequence to 38.2%. On the other hand, the percentage of preservice teachers holding constructivist preferences for dealing with student errors increased from 22.9% before the seminar sequence to 31.3% after it. The proportion of teachers providing insufficient information about how they would handle student errors increased from 26% in the pre-test to 30.5% in the post-test.

The changes in the preferences during the university seminar sequence are shown in Table 4.23. The cells display the number of participants who were classified in the preference class named on the row before the university seminar sequence and on the column after it. In this way, it can be observed in the dark-gray diagonal that 64 from 131 preservice teachers' preferences for dealing with student errors remained the same after the seminar sequence. A total of 25 participants, highlighted with light gray on the table, provided less evidence about their preferences for dealing with student errors in the post-test than in the pre-test. Middle-gray cells exhibit the number of preservice teachers who changed towards more instructivist-oriented preferences. From the 18 participants who changed in this direction, only five did it from constructivist preferences towards instructivist preferences. Conversely, 24 participants changed towards constructivist-oriented preferences, from which 18 had shown instructivist-oriented ways for dealing with errors before the university seminar sequence.

Summarizing these results, it is relevant to highlight the great proportion (48.9%) of participants who did not change their preference for dealing with student errors during the short university seminar sequence. It is also remarkable that far more participants moved from instructivist preferences to constructivist preferences than vice versa.

Table 4.23 Changes in the preferences for dealing with student errors from the pre-test to the post-test

		Class in the post-test			Total
		Instructivist preferences	Constructivist preferences	Missing evidence	
Class in the pre-test	Instructivist preferences	32	18	17	67
	Constructivist preferences	5	17	8	30
	Missing evidence	13	6	15	34
Total		50	41	40	131

4.2.3 Relationship between Changes in Both Competence Components

The results on investigating the relationship between the changes in the preferences for dealing with student errors and the changes in the competence to hypothesize about causes of student errors are summarized in Table 4.24. The four types of changes in the preferences described in the previous section are located on the rows and the means and standard deviations of the parameters of the hypothesizing-about-causes-of-student-errors competence for each group at each testing time are indicated on the columns.

Table 4.24 Descriptive statistics of the hypothesizing-about-causes competence at both testing times for preference-changing-pattern groups

	N	Pre-test		Post-test	
		Mean	Std. Deviation	Mean	Std. Deviation
No change	64	49.61	10.20	52.32	10.26
Less evidence	25	48.74	8.79	51.05	11.73
Towards instructivism	18	48.91	12.36	52.06	10.30
Towards constructivism	24	53.11	9.50	55.37	6.91
Total	131	49.99	10.14	52.60	10.03

An ANCOVA was run to determine the effect of the patterns of change in the preferences for dealing with errors on post-intervention hypothesizing-about-causes competence after controlling for pre-intervention hypothesizing-about-causes competence. There was homogeneity of regression slopes as the interaction term was not statistically significant, $F(3, 123) = .736, p = .532$. As assessed by Shapiro-Wilk's test ($p > .05$), standardized residuals were normally distributed for three of the change pattern groups and departed from a normal distribution for the group that changed towards instructivism. However, visual inspection of normal Q-Q plots revealed an approximately normal distribution for this group. There was homoscedasticity and homogeneity of variances, as assessed by visual inspection of a scatterplot and Levene's test of homogeneity of variance ($p = .079$), respectively. There was only one outlier in the data, as assessed by inspection of cases with standardized residuals greater than ± 3 standard deviations. After adjustment for pre-intervention hypothesizing-about-causes competence, there was not a statistically significant difference in post-intervention hypothesizing-about-causes competence between groups with different patterns of change on their preferences for dealing with student errors, $F(3,126) = .358, p = .783$. In other words, no significant relationship could be identified between changes in participants' preferences for dealing with students' errors and the development of their hypothesizing-about-causes competence.

4.2.4   Influence of Preservice Teachers' Characteristics on Changes in the Diagnostic Competence

Considering the diverse effect sizes of the changes that the university seminar sequence generated on the participants, the influence of several background characteristics on these changes was investigated. To do this, the same background characteristics from the cross-sectional analyses were used and associated both with the competence to hypothesize about causes of students' errors and with the preferences for dealing with errors. In the following, these associations are described.

Influence of Beliefs
In this section, the influence of preservice teachers' beliefs on the development of both components of the diagnostic competence will be examined. To explore the association between participants' beliefs about the nature of mathematics and about the learning of mathematics and the development of their competence to hypothesize about causes of student errors from the pre- to the

post-intervention testing time, a multiple regression analysis was conducted for each beliefs scale. Each analysis considered the post-intervention person parameter for the hypothesizing-about-causes competence as the dependent variable and two predictors, the pre-intervention person parameter for this competence and the ipsative values of the respective beliefs scale.

The multiple regression model that included the constructivist beliefs about the nature of mathematics statistically significantly predicted the post-intervention performance, $F(2,128) = 11.591, p = .000$, adj. $R^2 = .153$. However, the beliefs variable did not add significantly to the prediction ($\beta = 2.059, p = .176$). Similarly, the model including the variable with the constructivist beliefs about the learning of mathematics statistically significantly predicted the post-intervention competence to hypothesize about causes of student errors, $F(2,128) = 12.445, p = .000$, adj. $R^2 = .163$, but the beliefs variable did not contribute significantly to the model ($\beta = 2.616, p = .071$). This means that neither the constructivist beliefs about the learning of mathematics nor those about the nature of mathematics collected at the pre-intervention testing-time contribute to the prediction of the changes on the competence to hypothesize about causes of student errors after the university seminar sequence.

The association between the changes in the preferences for handling student errors and preservice teachers' beliefs is displayed in Table 4.25 and was explored using a one-way ANOVA. Results indicate that the means of the ipsative values for the constructivist beliefs about the nature of mathematics do not differ significantly between participants with different change patterns in their preferences for dealing with student errors, $F(3, 127) = 1.311, p = .274$. For the beliefs about the learning of mathematics a one-way Welch ANOVA was conducted because there was heterogeneity of variances. The adjusted F statistic suggests that the means of the ipsative values for the constructivist beliefs about the learning of mathematics do not differ significantly between participants with different change patterns in their preferences for handling student errors, Welch's $F(3, 47.033) = .698, p = .558$. In brief, no significant relationship was identified between changes in participants' preferences for dealing with students' errors and their beliefs about the nature of mathematics or their beliefs about the learning of mathematics.

Influence of Professional Knowledge
Knowledge about mathematics, its teaching and learning can facilitate the incorporation of new knowledge about errors, widen the perspectives to search for causes of students' errors, increase the repertoire of activities to handling these errors and give support to the ideas preservice teachers are able to suggest. For these reasons, the association of preservice teachers' professional knowledge and

Table 4.25 Descriptive statistics for beliefs about the nature of mathematics and beliefs about the learning of mathematics for preference-changing-pattern groups

	N	Nature of Mathematics as Inquiry Process			Learning of Mathematics as an active process		
		Mean	Std. Deviation	Std. Error	Mean	Std. Deviation	Std. Error
Less evidence	25	.31	.57	.11	1.27	.72	.14
Towards instructivism	18	.52	.66	.16	1.23	.84	.20
No change	64	.48	.58	.07	1.28	.59	.07
Towards constructivism	24	.50	.55	.11	1.42	.39	.08
Total	131	.46	.58	.05	1.30	.62	.05

the development of their diagnostic competence during the university seminar sequence was explored.

In order to study the association between preservice teachers' mathematical knowledge for teaching and their learning gains during the university seminar sequence regarding the competence to hypothesize about causes of student errors, a multiple regression analysis was conducted. It considered the post-intervention person parameters for the hypothesizing-about-causes-of-student-errors competence as the dependent variable and the pre-intervention person parameters for this competence and the mathematical knowledge for teaching as predictors. The model statistically significantly predicted the post-intervention performance, $F(2,128) = 16.018$, $p = .000$, adj. $R^2 = .200$. Mathematical knowledge for teaching contributed significantly to the prediction ($\beta = .258$, $p = .003$). This means that preservice teachers with higher scores in the professional knowledge test also show better gains in the development of their competence to hypothesize about causes of student errors. More specifically, an increase of one point in the mathematical knowledge for teaching person parameter is associated with an increase in the post-intervention hypothesizing competence parameter of 0.258, when the pre-intervention hypothesizing competence is held constant.

Table 4.26 displays the association between mathematical knowledge for teaching and the change patterns in the preferences for dealing with student errors. A one-way ANOVA was run to determine if the mathematical knowledge for teaching was distinctive for groups with different change patterns in their preferences for dealing with student errors. The differences between the four groups were not statistically significant, $F(3, 127) = .204$, $p = .894$. So, it can be stated that there is not a significant relation between preservice teachers' mathematical knowledge for teaching and the types of changes they experienced in their preferences for dealing with student errors during the university seminar sequence.

Table 4.26 Descriptive statistics for the mathematical knowledge for teaching for preference-changing-pattern groups

	N	Mean	Std. Deviation	Std. Error
Less evidence	25	49.56	9.45	1.89
Towards instructivism	18	49.99	8.54	2.01
No change	64	50.99	11.06	1.38
Towards constructivism	24	51.48	8.52	1.74
Total	131	50.67	9.93	.87

Influence of University Entrance Test Score

To investigate the influence that academic abilities may have on the development of the diagnostic competence during the university seminar sequence, the connection between the changes in each component of the diagnostic competence and preservice teachers' scores on their university entrance test was explored.

A multiple regression analysis was conducted to study the relation of the university entrance test scores with the changes in the competence to hypothesize about causes of student errors. The post-intervention person parameters of the competence were included as the dependent variable and the pre-intervention parameters of the competence, together with the university entrance test score, were added as predictors. The model statistically significantly predicted the post-intervention performance, $F(2,128) = 15.450$, $p = .000$, adj. $R^2 = .194$. Scores of the university entrance test contributed significantly to the prediction ($\beta = .051$, $p = .004$). This means that participants who entered university with higher scores show bigger improvements in the development of their competence to hypothesize about causes of student errors during the university seminar sequence. For instance, an increase of ten points in the university entrance test is associated with an increase in the post-intervention hypothesizing competence parameter of 0.510, when the pre-intervention hypothesizing competence is held constant. These results are in line with the TEDS-M study, which pointed out the high relevance of the university entrance examination score for future teachers' competencies at the end of their university studies (Blömeke, Kaiser, & Lehmann, 2010).

To assess the relationship between university entrance test scores and the change patterns in the preferences for dealing with student errors, a one-way Welch ANOVA was conducted because there was heterogeneity of variances. Table 4.27 summarizes the data of this association.

Table 4.27 Descriptive statistics for the university entrance test scores for preference-changing-pattern groups

	N	Mean	Std. Deviation	Std. Error
Less evidence	25	594.52	57.160	11.432
Towards instructivism	18	598.61	34.348	8.096
No change	64	611.64	51.437	6.430
Towards constructivism	24	618.54	26.909	5.493
Total	131	607.85	47.349	4.137

The adjusted F statistic suggests that the university entrance test mean scores do not differ significantly among the four preference change pattern groups, Welch's F (3, 52.408) = 2.043, $p = .119$. In other words, results suggest that there is not a significant relation between academic performance and the types of changes preservice teachers experienced on their preferences for dealing with student errors during the university seminar sequence.

Influence of Study Progress
To explore if preservice teachers who had completed more semesters of their teacher education program experience a greater improvement in their competence to hypothesize about causes of student errors, a multiple regression analysis was conducted. It considered the post-intervention competence parameters as the dependent variable and the number of completed semesters together with the pre-intervention competence parameters as the predictors. The model statistically significantly predicted the post-intervention performance, $F(2,128) = 18.094$, $p = .000$, adj. $R^2 = .220$. The number of completed semesters contributed significantly to the prediction ($\beta = 1.126$, $p = .000$). In other words, preservice teachers who were more advanced in their teacher education programs showed greater improvements in their competence to hypothesize about causes of student errors during the university seminar sequence. Each additional semester is associated with an increase of 1.13 in the post-intervention hypothesizing competence parameter when the pre-intervention hypothesizing competence is held constant.

The connection of the number of completed semesters with the change patterns of the preferences for dealing with student errors is shown in Table 4.28. A one-way Welch ANOVA was conducted to investigate if the number of passed semesters was significantly different among the four groups because the assumption of homogeneity of variances was not met.

Table 4.28 Descriptive statistics for the number of completed university semesters for preference-changing-pattern groups

	N	Mean	Std. Deviation	Std. Error
Less evidence	25	4.52	2.52	.50
Towards instructivism	18	5.89	3.38	.80
No change	64	6.11	2.28	.28
Towards constructivism	24	5.50	2.21	.45
Total	131	5.66	2.53	.22

Results indicate there is not a significant difference in the number of semesters students with different change patterns in their preferences for handling errors have completed, $F(3, 45.653) = 2.529$, $p = .069$. In other words, the changes experienced by preservice teachers in their preferences for dealing with student errors during the university seminar sequence occur regardless of the number of semesters they have completed in their teacher education programs.

Influence of Mathematics or Mathematics Education Courses

The number of mathematics or mathematics education courses completed by preservice teachers before participating in the study might help to connect the new knowledge and abilities to previous knowledge and therefore lead to greater improvements during the seminar sequence. To examine this relation, a multiple regression analysis was conducted considering the after the seminar sequence parameters of the competence to hypothesize about causes of student errors as the dependent variable and the pre-intervention competence together with the number of mathematics or mathematics education courses as predictors. The regression model significantly predicted preservice teachers' competence after the seminar sequence, $F(2,128) = 15.055$, $p = .000$, adj. $R^2 = .190$. The number of passed courses of mathematics or mathematics education contributed significantly to the prediction ($\beta = 1.049$, $p = .006$). Formulated differently, preservice teachers who have had more opportunities to learn within their teacher education programs, in the form of completed mathematics or mathematics education courses, experienced bigger gains in the development of their competence to hypothesize about causes of student errors during the university seminar sequence. If the pre-intervention hypothesizing competence is held constant, each additional mathematics or mathematics education university course is associated with an increase of 1.05 in the post-intervention parameter of the competence to hypothesize about causes of student errors.

Table 4.29 shows the association between the number of mathematics or mathematics education courses and the patterns of change in the preferences for dealing with student errors. In fact, the mean number of courses was higher in the group that did not experience any change in their preferences and in the one with the participants who changed towards more constructivist preferences.

Table 4.29 Descriptive statistics for the mathematics or mathematics education courses for preference-changing-pattern groups

	N	Mean	Std. Deviation	Std. Error
Less evidence	25	2.88	1.79	.36
Towards instructivism	18	3.39	2.70	.64
No change	64	4.19	2.02	.25
Towards constructivism	24	4.04	2.14	.44
Total	131	3.80	2.15	.19

However, a one-way ANOVA revealed that this difference is not statistically significant, $F(3, 127) = 2.645$, $p = .052$. Because the Shapiro-Wilk test indicated that the data were not normally distributed in the four groups ($p < .05$), a non-parametric test, the Kruskal-Wallis test, was conducted. It confirmed that the difference in the number of mathematics or mathematics education courses between the four groups was not statistically significant, X^2 (3) = 7.433, $p = .059$.

Influence of School Practicum
As previously acquired practical knowledge might have had an influence on the development of the diagnostic competence during the university seminar sequence, practical experiences such as school-based placements or practicum may also be an influential factor. The association between the number of school practicums preservice teachers have completed and their learning gains in the competence to hypothesize about causes of student errors during this study was evaluated using a multiple regression analysis. It considered the post-intervention

competence parameter as the dependent variable and the pre-intervention competence parameter and the number of school practicums as predictors. The regression model significantly predicted the post-intervention competence parameter, $F(2,128) = 17.771$, $p = .000$, adj. $R^2 = .217$. The number of school practicums in which preservice teachers had participated contributed significantly to the model ($\beta = 1.161$, $p = .001$). This means that preservice teachers who have completed more school placements benefited more from the university seminar sequence referred to in this study in that they showed greater gains in the development of their competence to hypothesize about causes of student errors. For instance, one additional school practicum experience (of one-semester duration) is associated with an increase of 1.61 in the post-intervention person parameter for the competence to hypothesize about causes of student errors, when the pre-intervention hypothesizing competence is held constant.

Regarding the association between the number of school practicums and the changes in the preferences preservice teachers exhibited to dealing with student errors, Table 4.30 shows that those who experienced no preference change had a higher mean of school practicums, whereas those who showed less evidence of their preference in the post-test had a lower mean of school practicums.

Table 4.30 Descriptive statistics for the number of school practicums for preference-changing-pattern groups

	N	Mean	Std. Deviation	Std. Error
Less evidence	25	2.92	2.75	.55
Towards instructivism	18	3.78	2.78	.66
No change	64	4.22	2.10	.26
Towards constructivism	24	3.13	2.37	.48
Total	131	2.92	2.75	.55

A one-way Welch ANOVA was used to evaluate if the differences among the change-pattern groups were statistically significant because there was heterogeneity of variances. The adjusted F statistic suggests that there is not a significant difference in the number of school practicums completed by participants with different change patterns in their preferences for dealing with student errors, $F(3, 44.883) = 2.280$, p $= .092$. Because the Shapiro-Wilk test suggested a non-normal distribution in the four groups, a non-parametric test was also run. The Kruskal-Wallis test also indicated that there is not a significant difference between the groups, $X^2 (3) = 5.821$, $p = .121$. Altogether, it can be stated that participants who experienced different types of changes in their preferences for dealing with student errors have not completed a significantly different number of one-semester school placements.

Influence of Teaching Experience in Primary Classrooms
To investigate further the influence that practical experiences working with primary school students may have on how preservice teachers acquire new knowledge and develop their diagnostic competence, it is of interest to examine the differences in the development of both competence components between preservice teachers who have already taught primary school students and those who have no teaching experience.

Table 4.31 displays the adjusted and unadjusted means and variability for post-intervention competence to hypothesize about causes of student errors. An ANCOVA was run to determine if the gains in the development of the competence to hypothesize about causes of student errors were significantly greater in the group of participants with experience teaching in primary classrooms than in the group without experience after controlling for the pre-intervention competence parameter. It can be seen from Table 4.31 that the adjusted mean was 4.85 higher for the group with experience teaching primary students. Indeed, results indicate that after adjustment for the pre-intervention competence parameter, there was a statistically significant difference in the post-intervention parameter of the competence to hypothesize about causes of student errors between both groups, $F(1, 128) = 7.314$, p $= .008$, partial $\eta^2 = .054$.

Table 4.31 Adjusted and unadjusted means and standard deviations for post-intervention competence to hypothesize about causes of student errors for participants with and without teaching experience in primary school

	Unadjusted			Adjusted	
	N	Mean	Std. Deviation	Mean	Std. Deviation
Without experience in primary teaching	38	47.86	9.64	49.13	1.49
With experience in primary teaching	93	54.50	9.38	53.98	.941

The association of teaching experience in the primary grades with the patterns of change in the preferences for dealing with student errors is exhibited in Table 4.32. A chi-square test of independence revealed a statistically significant relation, X^2 (3, $N = 131$) $= 12.466$, $p = .006$. The association is moderately strong (Cohen, 1988), Cramer's $V = .308$. From the analysis of the adjusted standardized residuals, it can be stated that both cells on the less evidence group and both cells on the group that did not experience any changes deviate significantly from independence, according to Agresti's (2007) guidelines. In the less-evidence group, there are more observed participants than expected without teaching experience and fewer participants than expected with experience teaching in the primary grades. The opposite is observed in the no-change group, fewer participants than expected have no experience teaching in the primary grades and there are more than expected with experience.

Table 4.32 Crosstabulation for preference-changing-pattern groups and teaching experience in primary grades

	N	Without experience in primary teaching	With experience in primary teaching
Less evidence	25	14	11
Towards instructivism	18	6	12
No change	64	12	52
Towards constructivism	24	6	18
Total	131	38	93

Influence of Mathematics Teaching Experience

In order to further explore the influence of teaching experience on the development of the diagnostic competence, it is relevant to study the association between the development of the competence and experience teaching specifically mathematics in primary classrooms. Table 4.33 shows the adjusted and unadjusted means and variability of the post-intervention competence to hypothesize about causes of student errors. It can be observed that preservice teachers who had taught mathematics in primary classrooms have a higher mean in the parameter than participants who have not had such experiences, even when adjusting for the pre-intervention parameter.

To determine if this difference in the competence growth was statistically significant, an ANCOVA was conducted. It indicates that there was a statistically significant difference in the post-intervention competence parameter between participants with and without experience teaching mathematics in primary classrooms, after controlling for the pre-intervention competence parameter, $F(1, 128) = 4.966$, $p = .028$, partial $\eta^2 = .037$.

Table 4.33 Adjusted and unadjusted means and standard deviations for post-intervention competence to hypothesize about causes of student errors for participants with and without experience teaching mathematics in primary school

	Unadjusted			Adjusted	
	N	Mean	Std. Deviation	Mean	Std. Error
Without experience in teaching mathematics	49	49.39	9.54	50.24	1.31
With experience in teaching mathematics	82	54.48	9.66	53.96	1.01

Similarly, there was also a statistically significant association of a medium-size effect between mathematics teaching experience and the patterns of change in the preferences for dealing with student errors, as revealed by the chi-square test of association, X^2 (3, $N = 131$) $= 10.403$, $p = .015$, Cramer's V $= .282$. Table 4.34 presents this relationship. The analysis of the adjusted standardized residuals reveals a similar situation to the results of the experience teaching in primary classrooms. In the less-evidence group, there are significantly more participants than expected without experience in mathematics teaching and fewer participants than expected who had taught mathematics. In addition, the contrary occurs in the group who did not demonstrate changes in their preferences for dealing with student errors. A significantly smaller than expected number of participants have not taught mathematics in primary classrooms and a greater than expected number of participants have taught.

Table 4.34 Crosstabulation for preference-changing-pattern groups and experience teaching mathematics in primary school

	N	Without experience teaching mathematics in primary	With experience teaching mathematics in primary
Less evidence	25	15	10
Towards instructivism	18	7	11
No change	64	16	48
Towards constructivism	24	11	13
Total	131	49	82

Influence of Private Tutoring Experience

Besides practical experiences that are part of the teaching education programs, such as school practicums and the teaching done within these activities, extracurricular teaching opportunities may also influence how preservice teachers acquire new knowledge. Private tutoring, in which preservice teachers support children individually or in very small groups after school, provide them with experiences to be in contact with students' thinking. These experiences may facilitate the identification of the relevance of the new knowledge being learned during the seminar sequence and the connection between previous and new knowledge and skills.

The means and variability of the post-intervention competence to hypothesize about causes of student errors for participants with and without private lesson experience (to students of any grade level) is displayed in Table 4.35.

Table 4.35 Adjusted and unadjusted means and standard deviations for post-intervention competence to hypothesize about causes of student errors for participants with and without private tutoring experience

		Unadjusted		Adjusted	
	N	Mean	Std. Deviation	Mean	Std. Error
Without private tutoring experience	47	51.58	11.79	52.06	1.35
With private tutoring experience	84	53.13	8.68	52.86	1.01

To evaluate the association between experience giving private lessons and the learning gains in the competence to hypothesize about causes of student errors during the university seminar sequence, an ANCOVA was conducted. Results suggest there is not a statistically significant difference in the hypothesizing-about-causes competence between preservice teachers with and without experience giving private tutoring to students of any grade level, after adjusting for the pre-intervention competence parameter, $F(1, 128) = .227, p = .635$.

An also non-significant result was obtained when assessing the differences in the competence to hypothesize about causes of student errors between participants with and without experience giving private tutoring to primary school students, whilst adjusting for the pre-intervention competence parameter, $F(1, 128) = .937, p = .335$. As can be seen in Table 4.36, preservice teachers who have tutored primary students performed better when hypothesizing about causes of student errors. However, their results were not significantly higher than those of preservice teachers who have never given private lessons to primary school students.

Table 4.36 Adjusted and unadjusted means and standard deviations for post-intervention competence to hypothesize about causes of student errors for participants with and without private tutoring experience in primary grades

		Unadjusted		Adjusted	
	N	Mean	Std. Deviation	Mean	Std. Error
Without primary-grades private tutoring experience	66	51.29	11.15	51.80	1.14
With primary-grades private tutoring experience	65	53.88	8.31	53.37	1.15

Likewise, the association between the patterns of change in the preferences for dealing with student errors and private tutoring experience was not statistically significant, according to the chi-square test of association, X^2 (3, $N = 131$) = 5.584, $p = .134$. As it can also be observed in Table 4.37, this means that preservice teachers with and without private tutoring experience are similarly distributed into the four preference groups.

Table 4.37 Crosstabulation for preference-changing-pattern groups and private tutoring experience

	N	Without private tutoring experience	With private tutoring experience
Less evidence	25	14	11
Towards instructivism	18	6	12
No change	64	19	45
Towards constructivism	24	8	16
Total	131	47	84

By contrast, when considering only private tutoring experiences to students of the same grade levels being included in the university seminar sequence and in the assessments, i.e., primary grade levels, statistically significant differences of a medium-size effect are found in how preservice teachers with and without private tutoring experience are distributed among patterns of change on the preferences for dealing with student errors, X^2 (3, $N = 131$) = 8.222, $p = .042$, Cramer's V = .251. From the exploration of Table 4.38 and its corresponding adjusted standardized residuals, it can be stated that in the less-evidence group there are significantly more participants than expected without experience and fewer participants than expected who had privately tutored primary grades students.

Table 4.38 Crosstabulation for preference-changing-pattern groups and primary-grades private tutoring experience

	N	Without primary-grades private tutoring experience	With primary-grades private tutoring experience
Less evidence	25	18	7
Towards instructivism	18	11	7
No change	64	28	36
Towards constructivism	24	9	15
Total	131	66	65

Influence of Sessions Attendance

Although no direct causal relation can be established, the association between attendance to the sessions of the university seminar sequence and the development of the diagnostic competence can be explored. To analyze the connection between higher attendance to the seminar sequence and changes in both facets of the diagnostic competence, participants were categorized into those who attended more or less than half of the sessions. A great proportion (90%) of the preservice teachers attended more than half of the sessions, only 13 participants attended either one or two sessions. Because of this difference in both group sizes, results must be interpreted with caution.

Table 4.39 displays the relationship between session attendance and the post-intervention competence to hypothesize about causes of student errors. Means and variability adjusted for the pre-intervention competence parameter are also included in the table.

After adjusting for the pre-intervention competence parameter, the post-intervention mean is higher in the group that attended 3–4 sessions. However, results of an ANCOVA indicate that the gains in the development of the competence to hypothesize about causes of student errors was not significantly greater in

Table 4.39 Adjusted and unadjusted means and standard deviations for post-intervention competence to hypothesize about causes of student errors according to sessions attendance

		Unadjusted		Adjusted	
	N	Mean	Std. Deviation	Mean	Std. Error
1–2 Sessions	13	53.29	7.90	51.27	2.60
3–4 Sessions	118	52.49	10.11	52.72	.85

the group of participants who attended 3–4 sessions than in the group who attended 1–2 sessions after controlling for the pre-intervention competence parameter, $F(1,128) = .278, p = .599$. This result must be carefully interpreted because the number of participants who attended only one or two sessions is very small.

The association between session attendance and the changes in the preferences for dealing with student errors is shown in Table 4.40. Results of a chi-square test of association confirm that there are statistically significant differences of a medium-size effect on how preservice teachers who attended more or less than a half of the sessions are distributed among patterns of change on the preferences for dealing with student errors, X^2 $(3, N = 131) = 8.079, p = .044$, Cramer's V $= .248$.

Table 4.40 Crosstabulation for preference-changing-pattern groups and sessions attendance

	N	1–2 Sessions	3–4 Sessions
Less evidence	25	1	24
Towards instructivism	18	2	16
No change	64	4	60
Towards constructivism	24	6	18
Total	131	13	118

Influence of the Level of Participation during the Sessions
Besides sessions' attendance, participants' active involvement in the activities during the university seminar sequence might also be relevant to the development of their diagnostic competence. After the university seminar sequence, participants self-reported their level of engagement into one of four categories: very active, active, little active or barely active. Although self-reported measures are

somehow problematic because of social desirability bias, response bias and different interpretations of similar situations, in this case, it enabled the collection of data about the level of participation at least in an explorative sense.

Table 4.41 exhibits the relationship between preservice teachers' self-reported level of participation and their unadjusted and adjusted parameters of the competence to hypothesize about causes of student errors. From the table, it is evident that preservice teachers who had a very active participation during the sessions also showed a high performance in the post-intervention test, even when controlling for the pre-intervention competence parameter.

Table 4.41 Adjusted and unadjusted means and standard deviations for post-intervention competence to hypothesize about causes of student errors according to level of participation during the sessions

	N	Unadjusted		Adjusted	
	N	Mean	Std. Deviation	Mean	Std. Error
Barely active	2	50.00	13.65	51.69	6.50
Little active	26	53.26	9.59	52.26	1.81
Active	86	51.68	9.33	51.92	.99
Very active	17	56.34	12.50	56.48	2.22

To test the significance of the association between active engagement during the sessions and the gains in the development of the competence to hypothesize about causes of student errors, an ANCOVA was conducted. After adjustment for the pre-intervention competence parameter, results indicate there is not a statistically significant difference in post-intervention competence between preservice teachers who reported different levels of participation during the seminar sessions, F (3,126) $= 1.189$, $p = .317$. Considering that four levels of participation may have been a too fine differentiation, data were dichotomized into a group who reported either a barely or little active participation and a second group with active or very active participation. Grouped in this way, differences in the competence parameters were also non-significant, F (1,128) $= .054$, $p = .817$. Thus, it is not possible to determine a relation between the self-reported level of engagement during the sessions and preservice teachers' gains in their competence to hypothesize about causes of student errors during the university seminar sequence.

The connection between preservice teachers' self-reported level of participation and their change patterns in their preferences for handling student errors is summarized in Table 4.42. From the table, it can be suggested that preservice teachers are distributed independently of their level of participation in the four groups of patterns of change in the preferences for dealing with student errors. This is confirmed by the chi-square test of association, X^2 $(9, N = 131) = 10.811$, $p = .289$. Similarly, if preservice teachers' level of participation is dichotomized as described above, its relation with their change pattern in the preferences for dealing with student errors is also not statistically significant, X^2 $(3, N = 131)$ $= 2.276$, $p = .517$. Once again, these results must be interpreted with caution because they rely on the self-reported perception of preservice teachers about their participation level and these perceptions may be related to other factors as well.

Table 4.42 Crosstabulation for preference-changing-pattern groups and level of participation during the sessions

	N	Barely active	Little active	Active	Very active
Less evidence	25	0	8	16	1
Towards instructivism	18	1	2	10	5
No change	64	1	12	44	7
Towards constructivism	24	0	4	16	4
Total	131	2	26	86	17

4.2.5 Regression Analyses

To further investigate the influence of the various independent variables on the development of the diagnostic competence from before the seminar sequence until after it, regression analyses were conducted. The association of the gains in the competence to hypothesize about causes of student errors and the predictor variables was studied by means of a multiple regression analysis. To evaluate the connection between preservice teachers' changes in their preferences for dealing with student errors and the independent variables, a multinomial logistic regression analysis was used.

In a first stage, a multiple regression analysis with the post-intervention parameter for the hypothesizing about causes of student errors competence as the criterion variable and with simultaneous entry of the predictor variables was conducted. Entered predictor variables were those that showed statistically significant results in the explorative analyses described in the previous sections and the pre-intervention competence parameter. From these variables, three had to be removed because they were causing multicollinearity problems. The offending variables were university entrance test scores, number of completed semesters and primary-school teaching experience. The other five independent variables were included in the model, namely the number of mathematics or mathematics education university courses, mathematical knowledge for teaching, experience teaching mathematics in primary classrooms, number of school practicums and the pre-intervention parameter of the competence to hypothesize about causes of student errors. The multiple regression model statistically significantly predicted the post-intervention competence to hypothesize about causes of student errors, $F(5, 125) = 8.585$, $p = .000$, adj. $R^2 = .226$. Regression coefficients and standard errors can be found in Table 4.43.

Table 4.43 Summary of Multiple Regression Analysis with all predictors

	B (Unstandardized Coefficients)	Std. Error of B	β (Standardized Coefficients)	Significance
Constant	24.832	4.968		.000
Pre-intervention competence	.280	.080	.283	.001
Experience teaching mathematics	−2.149	2.283	−.106	.348
Courses of mathematics or mathematics education	.418	.425	.091	.328
Number of school practicums	1.034	.440	.253	.020
Mathematical knowledge for teaching	.190	.091	.191	.038

From the five variables included into the model, only three added statistically significantly to the prediction. The greatest contribution was from the pre-intervention competence parameter, ($\beta = .28$, $p < .05$). In addition, the number of school practicums and mathematical knowledge for teaching contributed to the prediction of the post-intervention competence parameter. In other words, the competence parameter reached after the seminar sequence can be predicted not only by looking at the baseline from which each individual started (i.e., pre-intervention competence parameter) but also their practical experiences in schools and their mathematical knowledge for teaching contributed to the gains they showed.

To select the simplest model, all models that included the three predictors that were significant in the initial multiple regression analysis, were compared using the AIC criteria, as suggested by Fahrmeir and colleagues (2013). As shown in Table 4.44, some models included the other two independent variables, and some did not. The selected model included only the variables that made significant contributions in the initial analysis, i.e. the pre-intervention competence, the number of school practicums and mathematical knowledge for teaching. It significantly predicts the post-intervention competence to hypothesize about causes of student errors, $F(3, 127) = 13.820$, $p = .000$, and it explains 22.8% of the variance.

Estimated coefficients and standard errors for the selected model are displayed in Table 4.45. All three variables added significantly to the regression. The highest contribution was made by the pre-intervention competence parameter ($\beta = .28$). The number of school practicums and mathematical knowledge for teaching also contributed to the prediction with coefficients with at least a small size ($\beta = .23$ and $\beta = .19$, respectively). Besides indicating that preservice teachers who obtained higher competence parameters on the pre-intervention test also achieved higher scores after the intervention, results indicate that the higher the mathematical knowledge for teaching, the higher a preservice teacher scores on the hypothesizing-about-causes test when controlling for the other two predictors. Similarly, if pre-intervention competence parameter and mathematical knowledge for teaching are controlled for, the more school practicums preservice teachers have done, the higher their competence to hypothesize about causes of student errors after the intervention.

The relationship between the change patterns in the preferences for dealing with student errors and the predictor variables was modeled using a multinomial logistic regression analysis. As in the cross-sectional analysis, the purposeful selection approach was chosen (Hosmer et al., 2013). In an initial stage, all variables with a significance value smaller than 0.25 in the individual explorative analyses were included into a preliminary model. Variables not contributing at traditional significance levels in this model were then removed. In the following steps, variables were purposefully removed and re-entered according to the authors' suggestions, until a final model with four predictors was identified as the best model. Sessions attendance, experience teaching in primary classrooms, private tutoring experience to primary students and the number of school practicums

Table 4.44 Potential models of a final multiple regression analysis for the post-intervention competence to hypothesize about causes of student errors

Model	Pre-intervention competence	Experience teaching mathematics	Courses of mathematics or mathematics education	Number of school practicums	Mathematical knowledge for teaching	AIC	R2
1	x	x	x	x	x	572.730	.226
2	x	x		x	x	571.738	.226
3	x		x	x	x	571.655	.227
4	x			x	x	570.391	.228

Table 4.45 Summary of multiple regression analysis—parsimonious model—for changes in the competence to hypothesize about causes of student errors

	B (Unstandardized Coefficients)	Std. Error of B	β (Standardized Coefficients)	Significance
Constant	25.812	4.832		.000
Pre-intervention competence	.278	.080	.281	.001
Number of school practicums	.929	.334	.227	.006
Mathematical knowledge for teaching	.186	.084	.186	.029

were the selected independent variables. This final model significantly improved the null model to predict preservice teachers' changes in their preferences for dealing with student errors, X^2 (12, N = 131) = 29.438, p = .003. As revealed in Table 4.46, statistically significant contributions to the prediction were made only by teaching experience in primary classrooms.

Table 4.46 Likelihood ratio tests for final multinomial regression model for the change patterns of the preferences for dealing with student errors

	AIC of reduced model	−2 log likelihood of reduced model	Chi-square	df	Significance
Intercept	171.303	141.303	.000	0	.
Number of school practicums	171.502	147.502	6.200	3	.102
Sessions attendance	171.446	147.446	6.143	3	.105
Experience in primary teaching	173.487	149.487	8.184	3	.042
Primary-grades private tutoring experience	171.221	147.221	5.919	3	.116

Nevertheless, examination of the parameter estimates shown in Table 4.47 reveals a significant effect of other variables as well. Participants without teaching experience in primary classrooms are significantly more likely to show less evidence about their preferences for dealing with student errors after the intervention than changing towards constructivist-oriented preferences. Similarly, preservice teachers without private tutoring experience to primary-grades students are more likely to provide poor evidence after the intervention about their preferences for dealing with student errors than changing their preferences in a constructivist direction. Findings also suggest that participants who attended only one or two of the four sessions are less likely to be in the group who did not exhibit changes in their preferences than in the group that modified their preferences towards constructivism. Although these results do not go in the expected direction, they do not allow for much interpretation since the total number of preservice teachers who only attended one or two sessions is very low (N = 13). Finally, results show that for each additional school practicum, preservice teachers are less likely to be in the less evidence, in the towards instructivism or in the no change group rather than in the towards constructivism group. In other words, participants who have done more school practicums are more likely to be in the group that changed towards constructivist preferences during the seminar sequence than in any of the other three groups of patterns of change in the preferences for dealing with student errors.

4.2.6   Discussion of Longitudinal Results

The previous sections explored the development of the diagnostic competence during the university seminar sequence. The changes in both facets of the competence and their connection with other independent variables or predictors were investigated to examine the first and third hypotheses of this study (see chapter 2). As with the cross-sectional results, the Holm-Bonferroni correction needs to be applied at this stage to account for the problem of multiple testing. Again, the correction was applied separately for each facet of the diagnostic competence. Many of the relationships were not statistically significant after the adjustments. Henceforth these results will be discussed under the framework of the hypotheses of this study. However, generalizations cannot be made from these results. On the contrary, they need to be interpreted carefully because of the exploratory nature of the study design.

To investigate the second hypothesis, the advances made by preservice primary teachers on their competence to hypothesize about causes of student errors during

Table 4.47 Parameter estimates of the final multinomial regression model for the change patterns of the preferences for dealing with student errors

	Less evidence				Towards instructivism				No change			
	B	Std. Error of B	Sig.	β	B	Std. Error of B	Sig.	β	B	Std. Error of B	Sig.	β
Intercept	−2.77	1.11	.01		−2.56	1.15	.03		−.32	.79	.68	
Number of school practicums	.41	.20	.04	1.51	.40	.21	.05	1.50	.32	.16	.04	1.37
1–2 sessions attendance	−1.99	1.17	.09	.14	−.91	.92	.32	.40	−1.60	.72	.03	.20
Without experience in primary teaching	2.45	1.01	.02	11.61	1.57	1.07	.14	4.79	.62	.85	.46	1.86
Without primary-grades private tutoring experience	1.46	.69	.03	4.29	1.16	.70	.10	3.20	.54	.55	.33	1.71

the university seminar sequence were evaluated by comparing their pre- and post-intervention competence parameter. As in Heinrichs (2015) study with preservice secondary teachers, results indicated that post-intervention parameters were significantly better than those before the seminar sequence. Limitations of the size of the effect may be, as already pointed out by Heinrichs, related to the short duration of the university seminar sequence. Although the seminar sequences had the same duration in both studies, results in the present study were, despite small, of a bigger size than those in Heinrichs study. This may be related to the hypothesis she suggested for the limited effect in her study, namely the lack of direct connection between the errors worked on during the seminar sequence and those included in the tests, which sometimes even came from different mathematical areas. In the present study, all the errors used in the tests and during the seminar sessions belonged to the area of numeracy and operations. However, the transfer of knowledge and skills might have been still challenging because although in the same area, only the errors from one of the sessions were from a very similar topic than those in the test, i.e. included place value issues. In this sense, only few knowledge from the sessions could enlighten answers in the post-intervention test. Alternatively, skills learned during the seminar sequence had to be flexibly applied to new error situations. Evidently, this is also only a hypothesis, but it contributes to the suggestion made by Heinrichs (2015).

Regarding the preferences for dealing with student errors shown by preservice teachers, changes were evaluated by comparing the preference-classes into which each participant was categorized before and after the university seminar sequence. This yielded four patterns of change. Most preservice teachers did not exhibit change, i.e. they showed the same preferences before and after the university seminar sequence. The proportion of participants showing instructivist preferences decreased considerably and the percentage of preservice teachers holding constructivist preferences increased after the university seminar sequence.

As with the changes in the hypothesizing-about-causes-of-errors competence, the short duration of the university seminar sequence may explain the high percentage of students who did not show changes in their preferences for dealing with student mathematical errors. In fact, teacher practices or teaching styles are difficult to modify, so it is noteworthy that some participants did change their preferences after only four 90-minute sessions. The increase in the number of participants showing constructivist preferences and the decrease in the number of those showing instructivist preferences may have their reasons on characteristics of the university seminar sequence. Although constructivist principles were not considered explicitly, they were implicit across the seminar sequence. The focus of the sessions was on understanding students' thinking underlying the errors, finding possible causes and identifying a variety of pedagogical strategies for helping students to overcome the misconceptions. Even more, the idea itself of not ignoring errors and analyzing them as a way to understand student thinking and design pedagogical strategies that bring each learner to their zone of proximal development is in the heart of the constructivist learning theory. Moreover, the texts used and the teaching principles from the school curriculum were all based on the constructivist learning theory. In addition, the reflections and discussions about the errors, their causes and related pedagogical strategies were strongly influenced by constructivism. Together this may have led a number of students to change their preferences for dealing with student errors to an approach closer to constructivism and to move away from instructivism.

The last aspect to consider in relation to the first hypothesis is the association between the development of both facets of the diagnostic competence. Results show that the competence to hypothesize about causes of student errors developed positively and similarly in the four groups of patterns of change in the preferences for dealing with student errors. Therefore, it can be stated that there is no particular association between the way preservice teachers changed their preferences for dealing with student errors and the development of their competence to hypothesize about causes of errors.

To examine hypothesis number 3 and find influential factors, the relationships between a variety of background variables and the development of preservice teachers' diagnostic competence, operationalized onto both facets, were analyzed. Although many variables yielded significant results in their connections with the development of the competence to hypothesize about causes of student errors in the individual explorative analyses, only four remained significant after the Holm-Bonferroni adjustment. The advancement in the teacher education program (number of completed semesters), the number of school practicums, mathematical knowledge for teaching and the university entrance test score were relevant for the gains in the learning to find causes for student errors. The combination of these influential factors is interesting because it includes variables from different areas. Not only general academic abilities (operationalized as the university entrance test score) appear to be relevant, but also specialized knowledge for teaching mathematics contributes to having greater gains in the development of this competence. This reaffirms the relevance of specialized knowledge as a necessary tool to understand students' thinking, as a knowledge base on which (preservice) teachers can ground their interpretations of students' reasoning. On the other hand, the number of school practicums in which preservice teachers have had opportunities to observe and interact with primary school children and to observe and be involved in teaching played a significant role in the development of the competence. In addition, preservice teachers who had completed more semesters of their teacher education programs benefitted more from the short university seminar sequence. The higher number of completed semesters presumably implies that they have not only more specialized knowledge for teaching mathematics but also more general pedagogical knowledge and more experience in schools observing and teaching in primary classrooms.

Similarly, the regression analysis yielded as significant predictors the number of school practicums and mathematical knowledge for teaching when controlling for the pre-intervention competence to hypothesize about causes of student errors. Altogether, the corrected single analyses and the regression analysis highlight the complexity of the knowledge and experience required to interpret students' thinking and search for causes of their mathematical errors and confirms that it makes sense to locate this kind of opportunities to learn when preservice teachers have already had some practical experience and had acquired a first pedagogical and mathematical knowledge base.

Regarding the variables that may have an influence on the changes of the preferences to dealing with student errors, no factor remained significant after the Holm-Bonferroni correction. Although teaching experience in primary classrooms, experience teaching mathematics in primary classrooms, giving private lessons to primary students and attendance to the sessions of the university seminar sequence yielded significant associations in the single-variable analyses, they were no longer significant after the correction. Thus, results do not provide evidence about the reasons why some preservice teachers may have changed their preferences for dealing with student errors in any particular direction.

However, it is interesting to notice that three of the four variables that yielded significant results in the single-variable analyses were related to teaching experience. Moreover, the multinomial logistic regression model included, besides session attendance, only predictors related to practical experience, namely number of school practicums, teaching experience in primary classrooms and private tutoring to primary students. This insinuates that the types of changes preservice teachers undergo during the university seminar sequence may be related to their practical experiences. In particular, results suggest that the more experienced preservice teachers are, the more likely it is that they change towards constructivist preferences instead of towards instructivist preferences, giving less evidence or not changing at all. Again, these conclusions need to be interpreted with caution and taken only as hints that need to be further explored and confirmed in other studies because they come from the interpretation of results prior to the multiple-testing significance correction.

In conclusion, the third hypothesis could only be partially confirmed. On the one hand, some influential factors were found for the development of the hypothesizing about causes of errors facet of the competence, but on the other hand, no significant factors were found to explain the changes in preservice teachers' preferences for dealing with student errors.

Summary, Discussion and Outlook

5

In the present study, a university seminar sequence was developed with the aim of promoting the development of preservice primary school teachers' diagnostic competence in error situations within initial teacher education. Additionally, an online pre- and post-test assessment was designed to evaluate the development of the competence. Both the seminar sequence and the assessment instrument were based on Heinrichs' (2015) model of the diagnostic process in error situations. The model distinguishes three steps considered relevant in teaching situations in which students' errors arise, namely perceiving or identifying the error, developing hypotheses about reasons for the error, and then making instructional decisions for dealing with the error, so students overcome their misconception.

The seminar sequence consisted of four 90-minute sessions, in which preservice teachers participated in individual and collaborative analysis of samples of primary school students' work that showed mathematical errors. Future teachers worked through the three-phase diagnostic process several times and participated in productive discussions about student thinking and mathematics teaching and learning. The pre- and post-test assessment design included the collection of data about participants' background information, beliefs about the nature of mathematics and mathematics teaching and learning, their mathematical knowledge for teaching, and their diagnostic competence in error situations. The latter was evaluated with an online test developed for the purposes of the study and based on the three-phase diagnostic process model named above. Short videos were used to present the errors, and preservice teachers were asked to answer both open and closed items covering all three phases of the model.

A total of 131 preservice teachers from eleven Chilean universities, located in the capital city, participated in the sessions and answered the questionnaire. The sample consists of preservice teachers taking the university courses in which

M. Larrain Jory, *Preservice Primary Teachers' Diagnostic Competences in Mathematics*, Perspektiven der Mathematikdidaktik,
https://doi.org/10.1007/978-3-658-33824-4_5

the present study sessions were included and of a number who volunteered to participate in an additional course. Thus, the sample is not representative, and the results should be carefully interpreted.

The evaluation of the pre- and post-test items was carried out with qualitative and quantitative methods. The dichotomous items evaluating preservice teachers' competence to hypothesize about causes of students' errors were analyzed using Item Response Theory, providing an estimation of each participant's underlying latent competence. The open-response items evaluating preservice teachers' competence to deal with students' errors were first coded using evaluative qualitative text analysis. These codes were then used to extract classes of preferences for dealing with student errors with the aid of the latent class analysis method.

In a further analysis step, both components of preservice teachers' diagnostic competence in error situations were correlated with the participants' background and disposition characteristics. This allowed for a deeper understanding of the configuration of the diagnostic competence in the participating preservice teachers. In addition, the changes in the error-diagnostic competence from pre-test to post-test and their relation to background characteristics of the preservice teachers were examined to assess the effectiveness of the university seminar sequence.

In the following, the results of the analysis of the data from a cross-sectional perspective and then the changes from the pre- to the post-test are summarized and discussed. The study's limitations are then examined, the implications of the results are discussed, and starting points for further research opportunities are outlined.

5.1 Summary and Discussion of Results

Two competence facets were distinguished for examining and describing preservice primary school teachers' diagnostic competence in error situations and evaluating changes in its development. Both the competence to hypothesize about causes of student errors and the preferences for dealing with student errors were investigated.

The first facet, namely the competence to hypothesize about causes of student errors, refers to the competence that is necessary to identify hypotheses about possible causes for the mathematical errors made by students. Thus, respondents showing a high level of this competence facet are capable of identifying several causes for a student error and discerning between possible and not possible causes for a particular error pattern.

For the second facet, i.e., preservice teachers' preferences for dealing with students' errors, a distinction was made between different ways of dealing with student errors. Participants' responses were categorized into those showing a more constructivist approach to the error and those showing evidence of a more traditional or instructivist approach to dealing with error situations. The group exhibiting a constructivist approach was characterized by actively involving the student, suggesting targeted strategies that address the particular error and start further learning from what the student already knows and focusing more strongly on conceptual understanding than on the sole development of procedural skills. By contrast, the group displaying an instructivist approach suggested handling strategies that focused primarily on the teacher explaining the correct procedure and demonstrating the solution, focused to a lesser extent on addressing the particular error pattern and gave similar importance to procedural and conceptual understanding. In addition, a third group was extracted from the analysis, for which only limited evidence of the aspects under study could be found.

These two competence facets of the diagnostic competence in error situations were then analyzed. First, in the context of a cross-sectional investigation, connections between each of the facets and other teacher characteristics were examined. Then, in the longitudinal analysis, their changes occurring during the period in which the university seminar sequence was held were considered. These analyses will be briefly summarized below.

5.1.1 Cross-sectional Results

In the cross-sectional analysis, both facets of preservice teachers' diagnostic competence in error situations, i.e., the competence to hypothesize about causes of student errors and their preferences for dealing with those errors, were related to other background and disposition characteristics.

The first analyses exploring the relationship between the competence to hypothesize about causes of students' errors and preservice teachers' background and disposition characteristics provided support to some correlations that were suggested in the sub-hypotheses. In particular, constructivist beliefs about the nature of mathematics and about mathematics teaching and learning were found to be correlated with a higher competence to find causes for student errors in the sample of this study. This provides support for the idea that a more flexible view of the nature of mathematics, allowing to understand the subject beyond a set of fixed rules and procedures, permits taking several perspectives in the comprehension of a topic and, hence, in the understanding of student thinking. Similarly, from a constructivist stance on the learning of mathematics, efforts on interpreting students' errors and understanding their mathematical reasoning play a particularly relevant role so that such beliefs would equip teachers with a willingness to look for causes of student errors.

In addition, general academic ability, measured by the university-entrance-test score, and stronger mathematical knowledge for teaching were related to a higher competence to hypothesize about causes of student errors. Especially remarkable is the relation with preservice teachers' knowledge about mathematics teaching, as it can be argued that this provides a foundation for interpreting students' thinking. In other words, a solid knowledge base may support the identification of relevant details in students' work and the establishment of relations between them and aspects of mathematics teaching and learning that help to understand student thinking and interpret student errors from several perspectives. Stronger mathematical knowledge may provide crucial tools to differentiate between a possible cause for a particular error and another that is not possible in a specific situation.

Furthermore, evidence could be found that participants in this sample who had experience teaching in primary school classrooms showed greater competence to make hypotheses about causes of student errors. This points to a relation between practical experience and the development of the competence. It may be conjectured that experience in schools could have led preservice teachers to develop a better understanding of primary school students' thinking and the need to take a flexible stance to understand young students' ideas, which are often communicated incompletely.

Significant relations of this competence with other characteristics could also be found in the exploratory analyses with individual independent variables, such as the number of mathematics and mathematics education courses and the number of school practicums that preservice teachers had completed, their practical experience teaching mathematics in primary school classrooms and offering private tutoring to primary school students. However, none of these variables were significantly correlated with preservice teachers' competence to hypothesize about causes of student errors after the multiple testing correction was made. Hence, these correlations should not be interpreted as direct causal connections but can only be cautiously regarded as bivariate individual effects and should not be summarized. Nonetheless, they can be interpreted as associations that occurred in the present sample and can be first indicators of favorable conditions for developing this facet of the diagnostic competence in error situations that can be further explored in new studies.

Moreover, the fact that the correlations are in line with those expected serves to confirm the conceptualization of this facet of the competence. Because some of the correlations suggested in the hypotheses were, in fact, correct, it is possible to state in this respect that this way of measuring the competence of hypothesizing about the causes of student errors actually measures what is being sought to be evaluated.

In relation to the second facet of the competence, the preferences for dealing with student errors, three different groups were extracted with the aid of latent class analysis methods. The three classes were differentiated according to the focus of their suggested strategies on either promoting conceptual or procedural understanding, to the degree to which the strategies were tailored to the particular student error, and to the evidence they showed of an active-learner or a teacher-directed instructional approach. The first class was composed of more than one-third of the participating preservice teachers (35.3% of the participants in the pre-test) and showed instructivist preferences for dealing with student errors. This group showed high probabilities of taking a teacher-directed approach for dealing with student errors, with a fairly high probability of tailoring the strategies to the particular error situations and non-distinctive preferences for a focus on either conceptual or procedural understanding. The second group was somewhat smaller (30.6% of the participants in the pre-test) and exhibited constructivist preferences for dealing with student errors. They showed clear preferences for an active-learner approach, they strongly focused on promoting a conceptual understanding of mathematics, and showed the highest probability of providing targeted strategies that addressed the error in the particular conditions of the situation showed. The third class also comprised over one-third of the participants and included those preservice teachers who showed little evidence for all of the evaluated indicators because their responses were very negligible.

To evaluate the hypotheses, individual analyses were conducted to investigate the correlations between the preferences for dealing with student errors of the participating preservice primary school teachers and their background and disposition characteristics. Results indicated that preservice teachers with stronger constructivist beliefs about the nature of mathematics and about mathematics teaching and learning also showed more constructivist preferences for dealing with student errors. In the same way, preservice teachers viewing mathematics as a set of rules and procedures and teacher-directed beliefs about mathematics learning were significantly more likely to be classified into the group showing instructivist preferences for dealing with student errors. These correlations provide support to the classification of the participants to the classes of preferences for dealing with student errors as an acceptable construct that is closely related to the presumed influencing factors. However, it is surprising that after the multiple test correction and in the regression model, only the relation of the preferences for dealing with student errors and the beliefs about the nature of mathematics remains significant, and the association with the beliefs about mathematics teaching and learning does not prove to be significant. Actually, the scale of beliefs about mathematics teaching and learning is theoretically closer to the preferences for dealing with errors, and it was also expected to empirically show a stronger association. Thus, it would be interesting to confirm and explore this particular result in further studies.

Besides participants' beliefs, only the number of completed mathematics or mathematics education courses and the number of completed semesters showed significant relations to preservice teachers' preferences for dealing with student errors. Participants in more advanced stages of their teacher education programs were more likely to show constructivist preferences for dealing with student errors than to be classified in either of the other two classes. Similarly, preservice teachers who reported they had completed a higher number of mathematics and mathematics education courses were more likely to be classified into the constructivist preferences class than in the instructivist or missing evidence classes. It can be presumed that participants who have completed more courses, not only in mathematics but also in general education, have had more opportunities to learn about, understand and internalize constructivist teaching principles, so they use them when they plan how to respond in error situations. This, in turn, suggests that university courses do play a role in changing preservice teachers' approaches to dealing with student errors. However, this needs to be understood as a slight suggestion and not as a firm statement because it is an interpretation of changes over time made from cross-sectional data.

5.1.2   Longitudinal Results

The data were also longitudinally analyzed to investigate the changes in both facets of preservice teachers' diagnostic competence in error situations from the pre- to the post-test. Besides examining the development of the competence, these analyses were also used to measure the effectiveness of the university seminar sequence. Results indicated that significant changes were achieved in both facets of the diagnostic competence.

Regarding the competence to hypothesize about causes of student errors, a significant improvement was observed in the post-test parameters in relation to those shown by participants before the university seminar sequence. However, the size of the effect was small. This limited effect can be explained by the short duration of the university seminar sequence, which responded to the institutionally caused time restrictions in which the seminar sequence was applied. Additionally, the complexity of transferring the newly acquired skills to different error situations may have contributed to the small effect.

The examination of individual characteristics that may be associated with greater changes in the development of the competence to hypothesize about causes of student errors yielded many positive connections. After the multiple testing correction, a number of characteristics related to knowledge and practical experience were found to be significantly associated. Not only general academic ability but also mathematical knowledge for teaching were found to be related to a greater development of this facet of the competence. This supports the role of specialized professional knowledge for understanding and interpreting student thinking. Additionally, the fact that the number of school practicums was also associated with more gains in this facet suggests that preservice teachers' observations and interactions with primary school students helped them to contextualize and make sense of what was being learned in the university seminar sequence, leading to a more productive learning experience. Presumably including both stronger knowledge and more practical experiences, participants who were more advanced in their teacher education programs showed higher development of this facet of the competence as well, reaffirming the relevance of both knowledge and teaching experience again. In fact, the regression analysis model included the mathematical knowledge for teaching and the number of school practicum variables as well, contributing significantly to the prediction.

Concerning the changes in preservice teachers' preferences for dealing with student errors, it was found that most participants did not exhibit changes, remaining after the seminar sequence in the same preference class in which they were categorized before their participation in it. Nevertheless, some preservice teachers did change their preferences so that, after the seminar sequence, there were fewer participants showing instructivist preferences and more participants holding constructivist preferences for dealing with student errors. Considering that teaching practices, instructional styles, or approaches are usually difficult to change, the development observed in some of the participants can be positively interpreted as an opportunity to promote more constructive approaches to dealing with error situations in the classroom. It is possible that characteristics of the course's structure and content, as well as its underlying approach, may have influenced these changes. Furthermore, explicitly attending to student errors and considering them as an opportunity to understand students' misconceptions are ideas closely connected to constructivism.

The search for relations between individual variables and the patterns of change in preservice teachers' preferences for dealing with student errors pointed to significant correlations to several characteristics mainly related to practical experience. Besides being related to attending the sessions of the university seminar sequence, which is in line with the results by Heinrichs (2015), the changes observed in preservice teachers' preferences for dealing with student errors were found to be related to their experiences teaching as generalists in primary classrooms, teaching specifically mathematics in primary classrooms and giving private lessons to children in primary school age. In particular, preservice teachers with more practical experience tended to show changes towards constructivist preferences for dealing with errors after participating in the seminar sequence. However, none of these relations remained significant after the multiple-testing correction and only teaching experience as generalists remained a significant predictor in the regression model. Consequently, these results must be interpreted with caution and can only be considered as indications that would need to be confirmed in further studies.

Altogether these results suggest that it is possible to foster the development of teachers' diagnostic competence in error situations already in initial teacher education. For such an endeavor, developing awareness about the role of errors in teaching and learning and the relevance of diagnosing students' thinking for effective teaching and providing a framework to analyze student errors appear to be especially relevant.

5.2 Limitations of the Study and Further Prospects

Although the results provide interesting insights into preservice primary school teachers' diagnostic competence in error situations and its development, the limitations of the study also need to be discussed. Some of them—already delineated in the discussion of the results and in the summary section—were restrictions to the interpretation of results. Limitations of the present study can be found in its design and in the assessment instruments developed to evaluate the competence.

One limitation of the design of the study is related to the seminar sequence and its implementation. The university seminar sequence designed to develop preservice teachers' diagnostic competence in error situations had different status in the universities included: it was incorporated into the mandatory activities in some teacher education programs and offered as a voluntary extracurricular activity by another university. Thus, the present study is not a laboratory-based study, but a field study in which the intervention took place in natural settings. Although the university seminar sequence was conducted by the same person in all the groups following the same detailed lesson planning of the sessions and using exactly the same materials, it cannot be assumed that the conditions were absolutely identical in all groups. For instance, the time between the sessions was different among the groups. This affects the validity and generalizability of the results.

Moreover, because of the conditions in which the university seminar sequence was implemented, probability sampling techniques could not be used. Instead, the sample in this study is a convenience sample and may, therefore, not be representative of the entire population of preservice primary school teachers, not even Chilean ones. Consequently, the results cannot be considered as representative for preservice teachers in general and statements should only be made at the level of the sample, which is still appropriate for the exploratory nature of the study.

Another constraint of the present study lies in its design. As it is often the case in field studies, no control group and only an intervention group could be included. Thus, interpretation of the results and statements can only refer to the group of participants and no general direct causal relationships can be established between the content of the university seminar sequence and the changes in preservice teachers' diagnostic competence since the effects could have also been caused by other factors that could not be controlled.

Yet another limitation is related to the design of the tests for assessing diagnostic competence used in the study. Despite all the efforts to balance the characteristics and difficulties of the four error tasks used as prompts for the instrument, some differences remained and could have had a slight influence on the results. Further studies should consider introducing modifications or using other error examples. Moreover, the selected error tasks were all within the number and operations content area of elementary mathematics. Hence, the significance of the study is limited to error diagnosis in this area. Additional studies would be needed to explore preservice teachers' diagnostic competence in other content areas, such as geometry, early algebra, data and probability, or even other content areas within numeracy, such as fractions. It would be interesting to explore how the competence varies across content areas. Even more, preservice teachers' diagnostic competence in error situations in problem solving or modeling tasks could also be further explored in newly designed studies. In such studies, it would be interesting to investigate the differentiated role of mathematical knowledge and pedagogical content knowledge in the interpretation of errors that may show more complex characteristics.

Despite the advantages of latent class analysis methods, they also present some limitations. In particular, when the extracted latent classes are used to make further correlation analyses, as in the present study, a certain degree of uncertainty is introduced. Because individuals are categorized into a latent class according to their highest probability, for some cases the probabilities may not be close to 1 and, therefore, a certain degree of uncertainty remains in their classification. Nevertheless, a high entropy value and high classification probabilities were obtained in the present study, diminishing the degree of uncertainty.

An additional constraint is related to the background information collected. Although data about preservice teachers' practical experiences were gathered, more detailed information about their characteristics would have been useful to further clarify the relation between school and teaching experience and the development of diagnostic competence in error situations. Despite the efforts made to specify the types of practical school experiences participants had, further qualitative characterizations may contribute to explain the role they play on the development of the competence to hypothesize about causes of student errors and on the changes in the preferences for dealing with those errors.

Since this study was a more exploratory study and many different relationships were examined using hypothesis tests, it should be noted that errors may have occurred due to multiple testing. Thus, to evaluate the hypotheses again, they should be tested in further studies addressing the considerations made so far, such as using an experimental study design that includes a control group and/or selecting a representative sample.

Moreover, the longitudinal analysis of the present study considers only a pre- and post-test comparison that evaluates changes in the development of the competence in a short period of time. It would be desirable to examine to what extent the identified changes are stable over time by conducting a follow-up study that investigates the development of the competence after some years of teaching experience. Such a study would need to consider the inclusion of anchor test items so that the equivalence of test scores could be determined.

Notwithstanding the limitations mentioned above, the present study has provided evidence suggesting that it is possible to foster the development of preservice primary school teachers' diagnostic competence in error situations already during initial teacher education. Although the results are not generalizable due to characteristics of the study and its explorative nature, they provide valuable insights for the international discussion about preservice primary school teachers' diagnostic competence in error situations and its development due to the strong cultural influence of Europe in Chile. Results showed that, even with the implementation of a relatively brief university seminar sequence, changes could be generated in preservice teachers' competence to hypothesize about causes of student errors and their preferences for dealing with them in classroom situations. These results suggest that the introduction of error analysis in early stages of teacher education contributes to raising awareness of the opportunities errors bring into the mathematics learning process. Informing and educating preservice teachers about the role of errors and providing them with a framework to work with and reflect about errors and their own approaches to dealing with errors in teaching situations appear to be relevant aspects in this respect.

Due to the significant benefit of practical teaching experience and professional knowledge for the development of the competence to hypothesize about causes of students' errors, first efforts to develop preservice teachers' diagnostic competence appear to be particularly promising if they can be located in middle or advanced stages of initial teacher education when some professional knowledge has already been constructed and connections to practical experiences are available. Moreover, positive changes in preservice teachers' preferences for dealing with student errors are more likely to occur if they have already had practical teaching experiences. Again, this points to the need for incorporating diagnostic-competence learning opportunities in teacher education stages in which preservice teachers have already been through some practical school experiences. With many initial teacher education programs including school practicums all along the university years, it seems plausible to systematically integrate learning opportunities aimed at promoting the development of diagnostic competence. The present study provided suggestions on how such courses may be structured.

As one of the reasons attributed to the limited size of the effect of the seminar sequence on the development of the competence was its short duration, it would be desirable to design and implement longer seminar sequences. If more time were available, it would be possible to broaden the range of errors covered and deepen the pedagogical content knowledge and mathematical knowledge related to them. Analyzing and reflecting upon student errors, their causes and different ways to dealing with them in a wider variety of mathematical content and skills areas would probably lead to greater positive changes in preservice teachers diagnostic competence in error situations and to a stronger foundational base to the further development of the competence during their teaching career.

Altogether, results of the present study reveal that the development of preservice teachers' diagnostic competence in error situations is a complex phenomenon in which several factors interact. These factors, as suggested in the conceptualization of competence as a continuum by Blömeke et al. (2015), include affect-motivational aspects, such as beliefs, as well as professional knowledge. In addition, practical teaching experience appears to play a prominent role. This complexity implies great challenges for teacher educators, who need to provide opportunities to learn, addressing all these dimensions. Similarly, the mentioned complexity is both challenging and interesting for researchers seeking to understand this multifaceted competence and its promotion.

References

Abs, H. J. (2007). Überlegungen zur Modellierung diagnostischer Kompetenz bei Lehrerinnen und Lehrern. In M. Lüders (Eds.), *Forschung zur Lehrerbildung* (pp. 63–84). Münster: Waxmann.

Agresti, A. (2007). *An introduction to categorical data analysis* (2nd ed.). Hoboken, NJ: Wiley.

An, S., & Wu, Z. (2012). Enhancing mathematics teachers' knowledge of students' thinking from assessing and analyzing misconceptions in homework. *International Journal of Science and Mathematics Education, 10*(3), 717–753.

Anders, Y., Kunter, M., Brunner, M., Krauss, S., & Baumert, J. (2010). Diagnostische Fähigkeiten von Mathematiklehrkräften und ihre Auswirkungen auf die Leistungen ihrer Schülerinnen und Schüler. *Psychologie in Erziehung und Unterricht, 57*(3), 175–193. https://doi.org/10.2378/peu2010.art13d

Arias, P. & Villarroel, T. (2019). *Radiografía a las carreras de pedagogía y propuestas para maximizar el impacto de la ley de desarrollo profesional docente.* Acción Educar. https://accioneducar.cl/radiografia-a-las-carreras-de-pedagogia-y-propuestas-para-maximizar-el-impacto-de-la-ley-de-desarrollo-profesional-docente/

Artelt, C., & Gräsel, C. (2009). Diagnostische Kompetenz von Lehrkräften. *Zeitschrift für Pädagogische Psychologie, 23*(34), 157–160.

Ashlock, R. B. (2010). *Error patterns in computation: Using error patterns to help each student learn.* (10th ed.). Boston: Allyn & Bacon.

Ávalos, B., & Matus, C. (2010). La formación inicial docente en Chile desde una óptica internacional: Informe nacional del estudio internacional IEA TEDS-M. *Santiago de Chile: Ministerio de Educación.*

Baker, F. B., & Kim, S. H. (2017). *The basics of item response theory using R.* New York, NY: Springer.

Ball, D. L., & Bass, H. (2003). Toward a practice-based theory of mathematical knowledge for teaching. In B. Davis & E. Simmt (Eds.), *Proceedings of the 2002 annual meeting of the Canadian Mathematics Education Study Group* (pp. 3–14). Edmonton, Alberta, Canada: Canadian Mathematics Education Study Group.

Ball, D. L., Hill, H. C., & Bass, H. (2005, Fall). Knowing mathematics for teaching: Who knows mathematics well enough to teach third grade, and how can we decide? *American Educator,* 14–46.

Ball, D. L., Thames, M. H., & Phelps, G. (2008). Content knowledge for teaching: What makes it special? *Journal of Teacher Education,* 59, 389–407.

Baroody, A. J. & Hume, J. (1991). Meaningful Mathematics Instruction: The case of fractions. *Remedial and Special Education, 12*(3), 54–68.

Bartel, M. E., & Roth, J. (2017). Diagnostische Kompetenz von Lehramtsstudierenden fördern. In J. Leuders, T. Leuders, S. Prediger, & S. Ruwisch, (Eds.) *Mit Heterogenität im Mathematikunterricht umgehen lernen* (pp. 43–52). Wiesbaden: Springer Spektrum.

Bartell, T. G., Webel, C., Bowen, B., & Dyson, N. (2013). Prospective teacher learning: recognizing evidence of conceptual understanding. *Journal of Mathematics Teacher Education, 16*(1), 57–79.

Barth, C. & Henninger, M. (2012). Fostering the Diagnostic Competence of Teachers with Multimedia Training – A Promising Approach? In I. Deliyannis (Ed.), *Interactive Multimedia.* InTech: Rijeka, Croatia. doi: https://doi.org/10.5772/37297

Baumert, J., & Kunter, M. (2006). Stichwort: Professionelle Kompetenz von Lehrkräften. *Zeitschrift für Erziehungswissenschaft, 9*(4), 469–520.

Baumert J., & Kunter M. (2013) The COACTIV Model of Teachers' Professional Competence. In M. Kunter, J. Baumert, W. Blum, U. Klusmann, S. Krauss, M. Neubrand (Eds.) *Cognitive activation in the mathematics classroom and professional competence of teachers* (pp. 25–48). Springer, Boston, MA.

Baumert J., Kunter M. (2013) The Effect of Content Knowledge and Pedagogical Content Knowledge on Instructional Quality and Student Achievement. In M. Kunter, J. Baumert, W. Blum, U. Klusmann, S. Krauss, M. Neubrand (Eds.) *Cognitive activation in the mathematics classroom and professional competence of teachers* (pp. 175–205). Springer, Boston, MA.

Baumert, J., Kunter, M., Blum, W., Brunner, M., Voss, T., Jordan, A., Klusmann, U., Krauss, S., Neubrand, M., & Tsai, Y.-M. (2010). Teachers' Mathematical Knowledge, Cognitive Activation in the Classroom, and Student Progress. *American Educational Research Journal, 47*(1), 133–180. https://doi.org/10.3102/0002831209345157

Beck, E. (2008). *Adaptive Lehrkompetenz: Analyse und Struktur, Veränderbarkeit und Wirkung handlungssteuernden Lehrerwissens.* Münster: Waxmann.

Bergman, L. R., & Magnusson, D. (1997). A person-oriented approach in research on developmental psychopathology. *Development and psychopathology, 9*(2), 291–319.

Berliner, D.C. (2001). Learning about and learning from expert teachers. *International Journal of Educational Research, 35*(5), 463–482.

Binder, K., Krauss, S., Hilbert, S., Brunner, M., Anders, Y., & Kunter, M. (2018). Diagnostic Skills of Mathematics Teachers in the COACTIV study. In T. Leuders, K. Philipp & J. Leuders (Eds.) *Diagnostic competence of mathematics teachers* (pp. 33–53). Springer, Cham.

Biza, I., Nardi, E., & Zachariades, T. (2018). Competences of mathematics teachers in diagnosing teaching situations and offering feedback to students: Specificity, consistency and reification of pedagogical and mathematical discourses. In T. Leuders, K. Philipp & J. Leuders (Eds.) *Diagnostic Competence of Mathematics Teachers* (pp. 55–78). Springer, Cham.

Blömeke, S. (2012). Does greater teacher knowledge lead to student orientation? The relationship between teacher knowledge and teacher beliefs. In J. König (Ed.), *Teachers' pedagogical beliefs: Definition and operationalisation—Connections to knowledge and performance—Development and change* (pp. 15–35). Münster: Waxmann.

Blömeke, S., & Delaney, S. (2012). Assessment of teacher knowledge across countries: A review of the state of research. *ZDM – The International Journal on Mathematics Education, 44*(3), 223–247. https://doi.org/10.1007/s11858-012-0429-7

Blömeke, S., Gustafsson, J. E., & Shavelson, R. (2015). Beyond dichotomies: Competence viewed as a continuum. *Zeitschrift für Psychologie, 223*(1), 3–13. https://doi.org/10.1027/2151-2604/a000194.

Blömeke, S., & Kaiser, G. (2014). Theoretical framework, study design and main results of TEDS-M. In S. Blömeke, F.-J. Hsieh, G. Kaiser, & W. H. Schmidt (Eds.), *International perspectives on teacher knowledge, beliefs and opportunities to learn* (pp. 19–47). Springer, Dordrecht.

Blömeke, S., Kaiser, G., König, J., & Jentsch, A. (2020). Profiles of mathematics teachers' competence and their relation to instructional quality. *ZDM – The International Journal on Mathematics Education, 52*(2), 329–342. https://doi.org/10.1007/s11858-020-01128-y

Blömeke, S., Kaiser, G., & Lehmann, R. (Eds.). (2010). *TEDS-M 2008. Professionelle Kompetenz und Lerngelegenheiten angehender Primarstufenlehrkräfte im internationalen Vergleich.* Waxmann Verlag.

Blömeke, S., Suhl, U., Kaiser, G., & Döhrmann, M. (2012). Family background, entry selectivity and opportunities to learn: What matters in primary teacher education? An international comparison of fifteen countries. *Teaching and Teacher Education, 28*(1), 44–55.

Borasi, R. (1994). Capitalizing on errors as "springboards for inquiry": A teaching experiment. *Journal for research in mathematics education, 25*(2), 166–208.

Bray, W. S. (2011). A collective case study of the influence of teachers' beliefs and knowledge on error-handling practices during class discussion of mathematics. *Journal for research in mathematics education, 42*(1), 2–38.

Brandt, J., Ocken, A., & Selter, C. (2017). Diagnose und Förderung erleben und erlernen im Rahmen einer Großveranstaltung für Primarstufenstudierende. In J. Leuders, T. Leuders, S. Prediger, & S. Ruwisch, (Eds.) *Mit Heterogenität im Mathematikunterricht umgehen lernen* (pp. 53–64). Wiesbaden: Springer Spektrum.

Brodie, K. (2014). Learning about learner errors in professional learning communities. *Educational studies in mathematics, 85*(2), 221–239.

Brown, J. S., & Burton, R. R. (1978). Diagnostic models for procedural bugs in basic mathematical skills. *Cognitive science, 2*(2), 155–192.

Brown, J. S., & Van Lehn, K. (1980). Repair theory: A generative theory of bugs in procedural skills. *Cognitive science, 4*(4), 379–426.

Brunner, M., Anders, Y., Hachfeld, A. & Krauss, S. (2011). Diagnostische Fähigkeiten von Mathematiklehrkräften. In M. Kunter, J. Baumert, W. Blum, U. Klusmann, S. Krauss & M. Neubrand (Eds.), *Professionelle Kompetenz von Lehrkräften* (pp. 215–234). Münster: Waxmann.

Brühwiler, C. (2014). *Adaptive Lehrkompetenz und schulisches Lernen: Effekte handlungssteuernder Kognitionen von Lehrpersonen auf Unterrichtsprozesse und Lernergebnisse der Schülerinnen und Schüler* (Vol. 91). Waxmann Verlag.

Carpenter, T., Coburn, T., Reys, R., & Wilson, J. (1976). Notes from National Assessment: Addition and multiplication with fractions. *The Arithmetic Teacher, 23*(2), 137–142.

Cayton, G. A., & Brizuela, B. M. (2008). Relationships between children's external representations of number. *Proceedings of PME 32, 2*, 265–272.

Cebulski, L. A., & Bucher, B. (1986). Identification and remediation of children's subtraction errors: A comparison of practical approaches. *Journal of School Psychology, 24*(2), 163–180.

Chauraya, M., & Mashingaidze, S. (2017). In-service teachers' perceptions and interpretations of students' errors in mathematics. *International Journal for Mathematics Teaching and Learning, 18*(3), 273–292.

Chen, S. Y., Feng, Z., & Yi, X. (2017). A general introduction to adjustment for multiple comparisons. *Journal of thoracic disease, 9*(6), 1725–1729.

Chi, M. T. H. (2011). Theoretical perspectives, methodological approaches, and trend in the study of expertise. In Y. Li & G. Kaiser (Eds.), *Expertise in mathematics instruction: an international perspective* (pp. 17–39). New York, NY: Springer.

Choy, B. H., Thomas, M. O., & Yoon, C. (2017). The FOCUS framework: characterising productive noticing during lesson planning, delivery and review. In E.O. Schack, M. H. Fisher & J. A. Wilhelm (Eds.) *Teacher noticing: bridging and broadening perspectives, contexts, and frameworks* (pp. 445–466). Springer, Cham.

Clements, D. H., & Sarama, J. (2014). *Learning and teaching early math: The learning trajectories approach*. Routledge.

Collins, L. M., & Lanza, S. T. (2010). *Latent class and latent transition analysis: With applications in the social, behavioral, and health sciences* (Vol. 718). John Wiley & Sons.

Cohen, J. (1988). *Statistical power analysis for the behavioral sciences* (2nd ed.). New York, NY: Psychology Press.

Cohen, J. (1960). A coefficient of agreement for nominal scales. *Educational and psychological measurement, 20*(1), 37–46.

Cohen, L., Manion, L., & Morrison, K. (2007). *Research methods in education* (6th ed.). Oxon: Routledge.

Cooper, S. (2009). Preservice teachers' analysis of children's work to make instructional decisions. *School Science and Mathematics, 109*(6), 355–362.

Cox, L. S. (1974). *Analysis, classification, and frequency of systematic error computational patterns in the addition, subtraction, multiplication, and division vertical algorithms for grades 2–6 and special education classes*. Technical report ED092407. Kansas City: Kansas University Medical Center. ERIC Document Reproduction Service.

Cox, L. S. (1975). Systematic errors in the four vertical algorithms in normal and handicapped populations. *Journal for Research in Mathematics Education*, 202–220.

Cronbach, L. (1955). Processes affecting scores on "understanding of others" and "assumed similarity." *Psychological Bulletin, 52*(3), 177–193. doi: https://doi.org/10.1037/h0044919

Darling-Hammond, L. (2000). Teacher Quality and Student Achievement. *Education policy analysis archives, 8*(1), 1–44. doi: https://doi.org/10.14507/epaa.v8n1.2000.

DeMars, C. E. (2018). Classical Test Theory and Item Response Theory. In P. Irwing, T. Booth & D. J. Hughes (Eds.) *The Wiley Handbook of Psychometric Testing* (pp. 49–73). Wiley Blackwell. doi: https://doi.org/10.1002/9781118489772.ch2.

Depaepe, F., Verschaffel, L., & Kelchtermans, G. (2013). Pedagogical content knowledge: A systematic review of the way in which the concept has pervaded mathematics educational research. *Teaching and teacher education, 34,* 12–25.

Dimitrov, D. M., & Rumrill Jr, P. D. (2003). Pretest-posttest designs and measurement of change. *Work, 20*(2), 159–165.

Döhrmann, M., Kaiser, G., & Blömeke, S. (2014). The conceptualisation of mathematics competencies in the international study TEDS-M. In S. Blömeke, F.-J. Hsieh, G. Kaiser, & W. H. Schmidt (Eds.), *International perspectives on teacher knowledge, beliefs and opportunities to learn* (pp. 431–456). Springer, Dordrecht.

Döhrmann, M., Kaiser, G., & Blömeke, S. (2018). The conception of mathematics knowledge for teaching from an international perspective: The case of the TEDS-M study. In *How Chinese acquire and improve mathematics knowledge for teaching* (pp. 57–81). Brill Sense.

Dziak, J. J., Lanza, S. T., & Tan, X. (2014). Effect size, statistical power, and sample size requirements for the bootstrap likelihood ratio test in latent class analysis. *Structural equation modeling: a multidisciplinary journal, 21*(4), 534–552.

Edelenbos, P., & Kubanek-German, A. (2004). Teacher assessment: The concept of 'diagnostic competence'. *Language testing, 21*(3), 259–283.

Fahrmeir, L., Kneib, T., Lang, S., & Marx, B. (2013). *Regression: Models, methods and applications.* Berlin, Heidelberg: Springer.

Felbrich, A., Schmotz, C., & Kaiser, G. (2010). Überzeugungen angehender Primarstufenlehrkräfte im internationalen Vergleich. In S. Blömeke, G. Kaiser & R. Lehmann (Eds.), *TEDS-M 2008. Professionelle Kompetenz und Lerngelegenheiten angehender Primarstufenlehrkräfte im internationalen Vergleich* (pp. 297–325). Münster: Waxmann.

Feingold, A., Tiberio, S. S., & Capaldi, D. M. (2014). New approaches for examining associations with latent categorical variables: applications to substance abuse and aggression. *Psychology of Addictive Behaviors, 28*(1), 257–267.

Fiori, C., & Zuccheri, L. (2005). An experimental research on error patterns in written subtraction. *Educational Studies in Mathematics, 60*(3), 323–331.

Fischer, F., Kollar, I., Ufer, S., Sodian, B., Hussmann, H., Pekrun, R., et al. (2014). Scientific reasoning and argumentation: Advancing an interdisciplinary research agenda in education. *Frontline Learning Research, 5,* 28–45. https://doi.org/10.14786/flr.v2i2.96.

Fuson, K. C. (1990). Conceptual structures for multiunit numbers: Implications for learning and teaching multidigit addition, subtraction, and place value. *Cognition and instruction, 7*(4), 343–403.

García-Pérez, M. A., & Núñez-Antón, V. (2003). Cellwise Residual Analysis in Two-Way Contingency Tables. *Educational and Psychological Measurement, 63*(5), 825–839. https://doi.org/10.1177/0013164403251280

Gerster, H.-D. (1982). *Schülerfehler bei schriftlichen Rechenverfahren – Diagnose und Therapie.* Freiburg: Herder.

Goldsmith, L. T., & Seago, N. (2011). Using classroom artifacts to focus teachers' noticing: affordances and opportunities. In M. G. Sherin, V. R. Jacobs, & R. A. Philipp (Eds.), *Mathematics teacher noticing: Seeing through teachers' eyes* (pp. 3–13). New York: Routledge.

González López, M. J., Gómez, P., & Restrepo, Á. M. (2015). Usos del error en la enseñanza de las matemáticas. Error Uses in teaching mathematics. *Revista de educación, 370*, 71–95. DOI: https://doi.org/10.4438/1988-592X-RE-2015-370-297

Götze, D., Selter, C., & Zannetin, E. (2019). *Das KIRA-Buch: Kinder rechnen anders. Verstehen und fördern im Mathematikunterricht.* Hannover: Klett.

Granà, A., Lochy, A., Girelli, L., Seron, X., & Semenza, C. (2003). Transcoding zeros within complex numerals. *Neuropsychologia, 41*(12), 1611–1618.

Hansen, A., Drews, D., Dudgeon, J., Lawton, F., & Surtees, L. (2017). *Children's errors in mathematics.* London: SAGE, Learning Matters.

Hatcher, L. (2013). *Advanced statistics in research: Reading, understanding, and writing up data analysis results.* Shadow Finch Media, LLC.

Heinrichs, H. (2015). *Diagnostische Kompetenz von Mathematik-Lehramtsstudierenden: Messung und Förderung.* Wiesbaden: Springer.

Heinrichs, H., & Kaiser, G. (2018). Diagnostic competence for dealing with students' errors: Fostering diagnostic competence in error situations. In T. Leuders, K. Philipp & J. Leuders (Eds.) *Diagnostic competence of mathematics teachers* (pp. 79–94). Springer, Cham.

Heinze, A. (2004). Zum Umgang mit Fehlern im Unterrichtsgespräch der Sekundarstufe I. *Journal für Mathematik-Didaktik, 25*(3/4), 221–244.

Heinze, A. (2005). Mistake-Handling Activities in the Mathematics Classroom. In Chick, H. L. & Vincent, J. L. (Eds.). *Proceedings of the 29th Conference of the International Group for the Psychology of Mathematics Education,* Vol. 3, pp. 105–112. Melbourne: PME.

Heinze, A., & Reiss, K. (2007). Mistake-handling activities in the mathematics classroom: Effects of an in-service teacher training on students' performance in geometry. In J.H. Woo, H.C. Lew, K.S. Park & D.Y. Seo (Eds.). *Proceedings of the 31st Conference of the International Group for the Psychology of Mathematics Education* (Vol. 3, pp. 9–16). Seoul: PME.

Heitzmann, N., Seidel, T., Opitz, A., Hetmanek, A., Wecker, C., Fischer, M. R., et al. (2019). Facilitating diagnostic competences in simulations: a conceptual framework and a research agenda for medical and teacher education. *Frontline Learning Research, 7*(4), 1–24.

Helmke, A. (2003). *Unterrichtsqualität erfassen, bewerten, verbessern.* Seelze: Kallmeyer-sche Verlag.

Helmke, A. (2017). *Unterrichtsqualität und Lehrerprofessionalität: Diagnose, Evaluation und Verbesserung des Unterrichts.* Bobingen: Klett Kallmeyer.

Helmke, A., & Schrader, F. W. (1987). Interactional effects of instructional quality and teacher judgement accuracy on achievement. *Teaching and Teacher Education, 3*(2), 91–98.

Hickendorff, M., Edelsbrunner, P. A., McMullen, J., Schneider, M., & Trezise, K. (2018). Informative tools for characterizing individual differences in learning: latent class, latent profile, and latent transition analysis. *Learning and Individual Differences, 66*, 4–15.

Hill, H. C., Ball, D. L., & Schilling, S. G. (2008). Unpacking pedagogical content knowledge: Conceptualizing and measuring teachers' topic-specific knowledge of students. *Journal for Research in Mathematics Education, 39*(4), 372–400.

Hill, H. C., Blunk, M. L., Charalambous, C. Y., Lewis, J. M., Phelps, G. C., Sleep, L., & Ball, D. L. (2008). Mathematical knowledge for teaching and the mathematical quality of instruction: An exploratory study. *Cognition and instruction, 26*(4), 430–511.

Hill, H. C., & Charalambous, C. Y. (2012). Teacher knowledge, curriculum materials, and quality of instruction: Lessons learned and open issues. *Journal of Curriculum Studies*, 44(4), 559–576.

Hill, H. C., & Chin, M. (2018). Connections between teachers' knowledge of students, instruction, and achievement outcomes. *American Educational Research Journal*, 55(5), 1076–1112.

Hill, H. C., Rowan, B., & Ball, D. (2005). Effects of teachers' mathematical knowledge for teaching on student achievement. *American Educational Research Journal*, 42, 371–406.

Hill, H. C., Schilling, S. G., & Ball, D. L. (2004). Developing measures of teachers' mathematics knowledge for teaching. *The Elementary School Journal*, 105(1), 11–30.

Hoge, R. D. & Coladarci, T. (1989). Teacher-Based judgments of academic achievement: a review of literature. *Review of Educational Research*, 59(3), 297–313. doi: https://doi.org/10.3102/00346543059003297

Holmes, V. L., Miedema, C., Nieuwkoop, L., & Haugen, N. (2013). Data-driven intervention: correcting mathematics students' misconceptions, not mistakes. *The Mathematics Educator*, 23(1).

Hosmer Jr, D. W., Lemeshow, S., & Sturdivant, R. X. (2013). *Applied logistic regression* (3rd ed.). Hoboken, NJ: Wiley.

Hoth, J. (2016). *Situationsbezogene Diagnosekompetenz von Mathematiklehrkräften: Eine Vertiefungsstudie zur TEDS-Follow-Up-Studie*. Wiesbaden: Springer Spektrum.

Hoth, J., Döhrmann, M., Kaiser, G., Busse, A., König, J., & Blömeke, S. (2016). Diagnostic competence of primary school mathematics teachers during classroom situations. *ZDM – The International Journal on Mathematics Education*, 48(1-2), 41–53. https://doi.org/10.1007/s11858-016-0759-y

Hutcheson, G. & Hutcheson, G. (2011). Multinomial logistic regression. In L. Moutinho & G. Hutcheson (Eds.), *The SAGE dictionary of quantitative management research* (pp. 209–212). London: SAGE Publications Ltd doi: https://doi.org/10.4135/9781446251119.n63

Jacobs, V. R., Lamb, L. L., & Philipp, R. A. (2010). Professional noticing of children's mathematical thinking. *Journal for Research in Mathematics Education*, 41(2), 169–202.

Jacobs, V. R., Lamb, L. L., Philipp, R. A., & Schappelle, B. P. (2011). Deciding how to respond on the basis of children's understandings. In M. G. Sherin, V. R. Jacobs, & R. A. Philipp (Eds.), *Mathematics teacher noticing: Seeing through teachers' eyes* (pp. 97–116). New York: Routledge.

Jacobs, V. R., Lamb, L. C., Philipp, R., Schappelle, B., & Burke, A. (2007, April). *Professional noticing by elementary school teachers of mathematics*. Paper presented at the American Educational Research Association Annual Meeting. Chicago, IL.

Jacobs, V. R., & Philipp, R. A. (2004). Mathematical Thinking: Helping Prospective and Practicing Teachers Focus. *Teaching Children Mathematics*, 11(4), 194–201.

Jäger, R. S. (1999). Urteil und Entscheidung. In R. S. Jäger & F. Petermann (Eds.), *Psychologische Diagnostik: ein Lehrbuch* (pp. 449–455). Weinheim: Beltz Psychologie-Verlag-Union.

Jäger, R. S. (2006). Diagnostischer Prozess. In F. Petermann & M. Eid (Eds.), *Handbuch der Psychologischen Diagnostik* (pp. 89–96). Göttingen: Hogrefe Verlag.

Jäger, R. S. (2010). Diagnostische Kompetenz und Urteilsbildung als Element von Lehrprofessionalität. In O. Zlatkin-Troitschanskaia, K. Beck, D. Sembill, R. Nickolaus & R. Mulder (Eds.), *Lehrerprofessionalität* (pp. 105–116). Landau: Verl. Empirische Pädagogik.

Jan, S. L., & Shieh, G. (2014). Sample size determinations for Welch's test in one-way heteroscedastic ANOVA. *British Journal of Mathematical and Statistical Psychology, 67*(1), 72–93.

Jensen, S., & Gasteiger, H. (2019). „Ergänzen mit Erweitern" und „Abziehen mit Entbündeln"–Eine explorative Studie zu spezifischen Fehlern und zum Verständnis des Algorithmus. *Journal für Mathematik-Didaktik*, 1–33.

Jost, D., Erni, J., & Schmassmann, M. (1992). *Mit Fehlern muß gerechnet werden*. Zürich: Sabe-Verlag.

Kaiser, G., Busse, A., Hoth, J., König, J., & Blömeke, S. (2015). About the complexities of video-based assessments: Theoretical and methodological approaches to overcoming shortcomings of research on teachers' competence. *International Journal of Science and Mathematics Education, 13*(2), 369–387.

Kaiser, G., Blömeke, S., Busse, A., Döhrmann, M., & König, J. (2014). Professional knowledge of (prospective) mathematics teachers–its structure and development. In Liljedahl, P., Nicol, C., Oesterle, S., & Allan, D. (Eds.) *Proceedings of the Joint Meeting of PME 38 and PME-NA 36, Vol. 1* (pp. 35–50). Vancouver: Canada: PME.

Kaiser, G., Blömeke, S., König, J., Busse, A., Döhrmann, M., & Hoth, J. (2017). Professional competencies of (prospective) mathematics teachers—Cognitive versus situated approaches. *Educational Studies in Mathematics, 94*(2), 161–182. https://doi.org/10.1007/s10 649-016-9713-8

Kaiser, G., & König, J. (2019). Competence Measurement in (Mathematics) Teacher Education and Beyond: Implications for Policy. *Higher Education Policy, 32*(4), 597–615.

Kearney, M. W. (2017). Cramer's V. In M. R. Allen (Ed.), *The sage encyclopedia of communication research methods* (Vols. 1–4). Thousand Oaks, CA: SAGE Publications, Inc doi: https://doi.org/10.4135/9781483381411.

Keith, N. & Frese, M. (2008). Effectiveness of error management training: A meta-analysis. *Journal of Applied Psychology, 93*(1), 59–69.

Kleinbaum, D. G., & Klein, M. (2010). *Logistic regression. A self-learning text.* (3rd ed.). New York: Springer.

Klug, J., Bruder, S., Kelava, A., Spiel, C., & Schmitz, B. (2013). Diagnostic competence of teachers: A process model that accounts for diagnosing learning behavior tested by means of a case scenario. *Teaching and teacher education, 30*, 38–46.

KMK [Sekretariat der Kultusministerkonferenz] (2019). Standards für die Lehrerbildung: Bildungswissenschaften. Beschluss der Kultusministerkonferenz vom 16.12.2004 i. d. F. vom 16.05.2019. Bonn: KMK.

König, J. (2012). Teachers' pedagogical beliefs: Current and future research. In J. König (Ed.), *Teachers' pedagogical beliefs: Definition and operationalisation—Connections to knowledge and performance—Development and change* (pp. 7–13). Münster: Waxmann.

König, J. & Blömeke, S. (2010). Pädagogisches Wissen angehender Primarstufenlehrkräfte im internationalen Vergleich. In S. Blömeke, G. Kaiser & R. Lehmann (Eds.), *TEDS-M 2008 – Professionelle Kompetenz und Lerngelegenheiten angehender Primarstufenlehrkräfte im internationalen Vergleich* (pp. 275–296). Münster: Waxmann.

König, J., Blömeke, S., Klein, P., Suhl, U., Busse, A., & Kaiser, G. (2014). Is teachers' general pedagogical knowledge a premise for noticing and interpreting classroom situations? A video-based assessment approach. *Teaching and Teacher Education, 38*, 76–88. https:// doi.org/10.1016/j.tate.2013.11.004

König, J., Blömeke, S., Paine, L., Schmidt, W. H., & Hsieh, F. J. (2011). General pedagogical knowledge of future middle school teachers: On the complex ecology of teacher education in the United States, Germany, and Taiwan. *Journal of Teacher Education, 62*(2), 188–201.

König, J., & Pflanzl, B. (2016). Is teacher knowledge associated with performance? On the relationship between teachers' general pedagogical knowledge and instructional quality. *European Journal of Teacher Education, 39*(4), 419–436. doi: https://doi.org/10.1080/02619768.2016.1214128

Koeppen, K., Hartig, J., Klieme, E., & Leutner, D. (2008). Current issues in competence modeling and assessment. *Zeitschrift für Psychologie/Journal of Psychology, 216*(2), 61–73.

Krauss, S., Brunner, M., Kunter, M., Baumert, J., Blum, W., Neubrand, M., & Jordan, A. (2008). Pedagogical content knowledge and content knowledge of secondary mathematics teachers. *Journal of Educational Psychology*, 100(3), 716.

Krauss, S., Kunter, M., Brunner, M., Baumert, J., Blum, W., Neubrand, M., Jordan, A. & Löwen, K. (2004). *COACTIV: Professionswissen von Lehrkräften, kognitiv aktivierender Mathematikunterricht und die Entwicklung von mathematischer Kompetenz*. In J. Doll & M. Prenzel (Eds.), *Bildungsqualität von Schule: Lehrerprofessionalisierung, Unterrichtsentwicklung und Schülerforderung als Strategien der Qualitätsverbesserung* (pp. 31–53). Münster, Germany: Waxmann.

Kuckartz, U. (2014). *Qualitative text analysis: A guide to methods, practice and using software*. London: Sage.

Kuckartz, U. (2019). Qualitative Text Analysis: A Systematic Approach. In G. Kaiser & N. Presmeg (Eds.), *Compendium for Early Career Researchers in Mathematics Education* (pp. 181–197). Cham: Springer.

Kühnhold, K., & Padberg, F. (1986). Über typische Schülerfehler bei der schriftlichen Subtraktion natürlicher Zahlen. *Der Mathematikunterricht, 32*(3), 6–16.

Lakens, D. (2013). Calculating and reporting effect sizes to facilitate cumulative science: a practical primer for t-tests and ANOVAs. *Frontiers in psychology, 4,* 863.

Lanza, S. T., & Rhoades, B. L. (2013). Latent class analysis: an alternative perspective on subgroup analysis in prevention and treatment. *Prevention Science, 14*(2), 157–168.

Larrain, M. (2016). Comprensión del razonamiento matemático de los estudiantes: una práctica pedagógica inclusiva. *UNIÓN: Revista Iberoamericana de Educación Matemática.* 45, 152–161.

Lenz, K., Dreher, A., Holzäpfel, L., & Wittmann, G. (2020). Are conceptual knowledge and procedural knowledge empirically separable? The case of fractions. *British Journal of Educational Psychology, 90*(3), 809–829.

Lenz, K., Holzäpfel, L. & Wittmann, G. (2019). Aufgaben als Lerngelegenheiten für konzeptuelles und prozedurales Wissen zu Brüchen – Eine vergleichende Schulbuchanalyse. *Mathematica Didactica 42(1)*. https://doi.org/10.1111/bjep.12333

Lerner, D., Sadovsky, P., & Wolman, S. (1997). El sistema de numeración: un problema didáctico. In C. Parra & I. Saiz. (Eds.). *Didáctica de las matemáticas. Aportes y reflexiones.* Buenos Aires: Paidós.

Leuders, T., Dörfler, T., Leuders, J., & Philipp, K. (2018). Diagnostic competence of mathematics teachers: Unpacking a complex construct. In T. Leuders, K. Philipp & J. Leuders (Eds.) *Diagnostic competence of mathematics teachers* (pp. 3–31). Springer, Cham.

Li, Y., & Kaiser, G. (2011). Expertise in mathematics instruction: Advancing research and practice from an international perspective. In Y. Li, & G. Kaiser (Eds.) *Expertise in mathematics instruction* (pp. 3–15). Springer, Boston, MA.

Linacre, J. M. (2002). What do infit and outfit, mean-square and standardized mean. *Rasch Measurement Transactions*, 16(2), 878.

Lucchini, G., Cuadrado, B. & Tapia, L. (2006). *Errar no es siempre un error.* Santiago, Chile: Fundación Educacional Arauco (Fundar). Retrieved from https://fundacionarauco.cl/wp-content/uploads/2018/07/file_3878_errar-no-es-siempre-un-error-1.pdf

MacDonald, P. L., & Gardner, R. C. (2000). Type I Error Rate Comparisons of Post Hoc Procedures for I j Chi-Square Tables. *Educational and Psychological Measurement*, 60(5), 735–754. https://doi.org/10.1177/00131640021970871

Malle, G. (1993). *Didaktische Probleme der elementaren Algebra.* Braunschweig: Vieweg.

Martínez, F., Martínez, S., Ramírez, H. & Varas, M.L. (2014). *Refip Matemática. Recursos para la formación inicial de profesores de educación básica. Proyecto FONDEF D09I-1023.* Santiago, Chile: SM. Available online at https://refip.cmm.uchile.cl/files/memoria.pdf

Mayring, P. (2000). Qualitative Content Analysis. *Forum Qualitative Sozialforschung/Forum: Qualitative Social Research, 1*(2). doi: https://doi.org/10.17169/fqs-1.2.1089

Mayring, P. (2015). Qualitative content analysis: Theoretical background and procedures. In A. Bikner-Ahsbahs, C. Knipping & N. Presmeg (Eds.), *Approaches to qualitative research in mathematics education* (pp. 365–380). Dordrecht: Springer.

McDuffie, A. R., Foote, M. Q., Bolson, C., Turner, E. E., Aguirre, J. M., Bartell, T. G., Drake, C. & Land, T. (2014). Using video analysis to support prospective K-8 teachers' noticing of students' multiple mathematical knowledge bases. *Journal of Mathematics Teacher Education, 17*(3), 245–270.

McGuire, P. (2013). Using online error analysis items to support preservice teachers' pedagogical content knowledge in mathematics. *Contemporary Issues in Technology and Teacher Education, 13*(3), 207–218.

McHugh, M. L. (2013). The chi-square test of independence. *Biochemia medica, 23*(2), 143–149.

McHugh M. L. (2012). Interrater reliability: the kappa statistic. *Biochemia medica, 22*(3), 276–282.

Mertens, W., Pugliese, A., & Recker, J. (2017). *Quantitative data analysis. A companion for Accounting and Information Systems Research.* Switzerland: Springer International Pubishing Switzerland. https://doi.org/10.1007/978-3-319-42700-3.

Meyer, J. P. (2014). *Applied measurement with jMetrik.* Oxon: Routledge.

Mineduc (2012). *Matemática. Programa de Estudio para tercer Año Básico.* Santiago, Chile: Author.

Moeller, K., Pixner, S., Zuber, J., Kaufmann, L., & Nuerk, H. C. (2011). Early place-value understanding as a precursor for later arithmetic performance—A longitudinal study on numerical development. *Research in developmental disabilities, 32*(5), 1837–1851.

Moser Opitz, E. & Nührenbörger, M. (2015). Diagnostik und Leistungsbeurteilung. In: R. Bruder, L. Hefendehl-Hebeker, B. Schmidt-Thieme, H-G. Weigand (Eds.) *Handbuch der Mathematikdidaktik.* Berlin, Heidelberg: Springer Spektrum.

Müller, C., Felbrich, A. & Blömeke, S. (2008). Überzeugungen zum Lehren und Lernen von Mathematik. In S. Blömeke, G. Kaiser & R. Lehmann (Eds.), *Professionelle Kompetenz angehender Lehrerinnen und Lehrer* (pp. 248–276). Münster: Waxmann.

Muijs, D. (2004). *Doing quantitative research in education with SPSS*. London: Sage.

Muthén, L. K., & Muthén, B. O. (1998–2015). *Mplus User's Guide: Statistical Analysis with Latent Variables.* (7th ed.). Los Angeles, CA: Muthén & Muthén.

National Council of Teachers of Mathematics. (2000) *Principles and standards for school mathematics.* Reston, VA: NCTM.

Nesher, P. (1987). Towards an instructional theory: The role of student's misconceptions. *For the learning of mathematics, 7*(3), 33–40.

Neumann, I., Neumann, K., & Nehm, R. (2011). Evaluating instrument quality in science education: Rasch-based analyses of a nature of science test. *International Journal of Science Education, 33*(10), 1373–1405.

Nolte, M. (2021). Fehler und ihre Bedeutung im mathematischen Lernprozess. *Lernen und Lernstörungen, 10*(1), 3–12.

Nylund, K. L., Asparouhov, T., & Muthén, B. O. (2007). Deciding on the number of classes in latent class analysis and growth mixture modeling: A Monte Carlo simulation study. *Structural equation modeling: A multidisciplinary Journal, 14*(4), 535–569.

O'Connell, N. S., Dai, L., Jiang, Y., Speiser, J. L., Ward, R., Wei, W., Carroll, R. & Gebregziabher, M. (2017). Methods for analysis of pre-post data in clinical research: a comparison of five common methods. *Journal of biometrics & biostatistics, 8*(1), 1–8.

OECD (2009). Teaching practices, Teachers' beliefs and attitudes in *Creating Effective Teaching and Learning Environments: First Results from TALIS.* OECD Publishing. doi: https://doi.org/10.1787/9789264068780-6-en

Oser, F., Hascher, T., & Spychiger, M. (1999). Lernen aus Fehlern: Zur Psychologie des "negativen Wissens". In W. Althof (Ed.) *Fehlerwelten: vom Fehlermachen und Lernen aus Fehlern* (pp. 11–41). Springer Fachmedien, Wiesbaden.

Ostermann, A. (2018). Factors Influencing the Accuracy of Diagnostic Judgments. In T. Leuders, K. Philipp & J. Leuders (Eds.). *Diagnostic Competence of Mathematics Teachers* (pp. 95–108). Springer, Cham.

Orozco-Hormaza, M., Guerrero-López, D. F., & Otálora, Y. (2007). Los errores sintácticos al escribir numerales en rango superior. *Infancia y aprendizaje, 30*(2), 147–162.

Padberg, F. (2002). *Didaktik der Bruchrechnung.* Spektrum, Heidelberg.

Padberg, F., & Benz, C. (2011). *Didaktik der Arithmetik. für Lehrerausbildung und Lehrerfortbildung.* 4th ed. Heidelberg: Spektrum.

Pankow, L., Kaiser, G., Busse, A., König, J., Blömeke, S., Hoth, J., & Döhrmann, M. (2016). Early career teachers' ability to focus on typical students errors in relation to the complexity of a mathematical topic. *ZDM – The International Journal on Mathematics Education, 48*(1-2), 55–67.

Pankow, L., Kaiser, G., König, J., & Blömeke, S. (2018). Perception of student errors under time limitation: are teachers faster than mathematicians or students?. *ZDM – The International Journal on Mathematics Education, 50*(4), 631–642.

Peng, A., & Luo, Z. (2009). A framework for examining mathematics teacher knowledge as used in error analysis. *For the learning of mathematics, 29*(3), 22–25.

Peterson, P., & Clark, C. M. (1978). Teachers' reports of their cognitive processes during teaching. *American Educational Research Journal, 15*, 555–565.

Philipp, K. (2018). Diagnostic competences of mathematics teachers with a view to processes and knowledge resources. In T. Leuders, K. Philipp & J. Leuders (Eds.) *Diagnostic competence of mathematics teachers* (pp. 109–127). Springer, Cham.

Porter, A. (1989). A curriculum out of balance: The case of elementary school mathematics. *Educational researcher, 18*(5), 9–15.

Power, R. J. D., & Dal Martello, M. F. (1990). The dictation of Italian numerals. *Language and Cognitive processes, 5*(3), 237–254.

Praetorius, A. K., Berner, V. D., Zeinz, H., Scheunpflug, A., & Dresel, M. (2013). Judgment confidence and judgment accuracy of teachers in judging self-concepts of students. *The Journal of Educational Research, 106*(1), 64–76.

Praetorius, A.-K., Lipowsky, F. & Karst, K. (2012). Diagnostische Kompetenz von Lehrkräften: Aktueller Forschungsstand, unterrichtspraktische Umsetzbarkeit und Bedeutung für den Unterricht. In R. Lazarides (Eds.), *Differenzierung im mathematisch-naturwissenschaftlichen Unterricht* (pp. 115–146). Bad Heilbrunn: Klinkhardt.

Prediger, S. (2010). How to develop mathematics-for-teaching and for understanding: The case of meanings of the equal sign. *Journal of Mathematics Teacher Education, 13*(1), 73–93.

Prediger, S., & Wittmann, G. (2009). Aus Fehlern lernen–(wie) ist das möglich. *Praxis der Mathematik in der Schule, 51*(3), 1–8.

Rach, S., Ufer, S., & Heinze, A. (2013). Learning from Errors: Effects of Teachers Training on Students' Attitudes towards and Their Individual Use of Errors. *PNA, 8*(1), 21–30.

Radatz, H. (1979). Error analysis in mathematics education. *Journal for research in mathematics education*, 163–172.

Radatz, H. (1980). *Fehleranalysen im Mathematikunterricht*. Braunschweig: Vieweg.

Rasch, G. (1977). On Specific Objectivity: An attempt at Formalizing the Request for Generality and Validity of Scientific Statements. Danish Yearbook of Philosophy, 14, 58–94.

Reinhold, S. (2015). Uncovering facets of interpreting in diagnostic strategies pre-service teachers use in one-on-one interviews with first-graders. In K. Krainer & N. Vondrová (Eds.), *Proceedings of the Ninth Congress of the European Society for Research in Mathematics Education* (pp. 2895–2901). Prague, Czech Republic: ERME.

Reisman, F. K. (1982). *A guide to the diagnostic of teaching of arithmetic* (3rd Ed.). Columbus: C. E. Merrill.

Reiss, K. & Hammer, C. (2013). *Grundlagen der Mathematikdidaktik: Eine Einführung für die Sekundarstufe*. Basel: Birkhäuser

Rheinberg, F. (1978). Gefahren pädagogischer Diagnostik. In K. J. Klauer (Ed.). *Handbuch der pädagogischen Diagnostik* (pp. 27–38). Düsseldorf: Schwann.

Riccomini, P. J. (2005). Identification and remediation of systematic error patterns in subtraction. *Learning Disability Quarterly, 28*(3), 233–242.

Richardson, V. (1996). The role of attitudes and beliefs in learning to teach. In J. Sikla, T. Buttery, & E. Guyton (Eds.) *Handbook of research on teacher education* (2nd ed., pp. 102–119). New York: Macmillan.

Rico, L. (1995). Errores y dificultades en el aprendizaje de las matemáticas. In J. Kilpatrick, P. Gómez, L. Rico (Eds.), *Educación matemática: errores y dificultades de los estudiantes. Resolución de problemas. Evaluación. Historia: Primer Simposio Internacional de Educación Matemática.* (pp. 69–108). Grupo Editorial Iberoamericana.

Rittle-Johnson, B., & Alibali, M. W. (1999). Conceptual and procedural knowledge of mathematics: Does one lead to the other?. *Journal of educational psychology*, 91(1), 175.

Rittle-Johnson, B., & Koedinger, K. (2009). Iterating between lessons on concepts and procedures can improve mathematics knowledge. *British Journal of Educational Psychology*, 79(3), 483–500.

Rittle-Johnson, B., Schneider, M., & Star, J. R. (2015). Not a one-way street: Bidirectional relations between procedural and conceptual knowledge of mathematics. *Educational Psychology Review*, 27(4), 587–597.

Rittle-Johnson, B., Siegler, R. S., & Alibali, M. W. (2001). Developing conceptual understanding and procedural skill in mathematics: An iterative process. *Journal of educational psychology*, 93(2), 346–362.

Roa, R. (2001). Algoritmos de cálculo. In E. Castro (Eds.) *Didáctica de las matemáticas en la educación primaria*, 231–254.

Robitzsch, A., Kiefer, T., & Wu, M. (2017). *TAM: Test analysis modules*. R package version 2.8-21. https://CRAN.R-project.org/package=TAM

Sánchez, A. B., & López Fernández, R. (2011). La transferencia de aprendizaje algorítmico y el origen de los errores en la sustracción. *Revista de educación*, 354, 429–445.

Sánchez-Matamoros, G., Fernández, C., & Llinares, S. (2019). Relationships among prospective secondary mathematics teachers' skills of attending, interpreting and responding to students' understanding. *Educational Studies in Mathematics*, 100(1), 83–99.

Santagata, R. (2005). Practices and beliefs in mistake-handling activities: A video study of Italian and US mathematics lessons. *Teaching and Teacher Education*, 21(5), 491–508.

Santagata, R. (2011). From teacher noticing to a framework for analyzing and improving classroom lessons. In M. G. Sherin, V. R. Jacobs, & R. A. Philipp (Eds.), *Mathematics teacher noticing: Seeing through teachers' eyes* (pp. 97–116). New York: Routledge.

Santagata, R., & Guarino, J. (2011). Using video to teach future teachers to learn from teaching. *ZDM – The International Journal on Mathematics Education*, 43(1), 133–145.

Santagata, R., & Lee, J. (2019). Mathematical knowledge for teaching and the mathematical quality of instruction: a study of novice elementary school teachers. *Journal of Mathematics Teacher Education*. https://doi.org/10.1007/s10857-019-09447-y

Santagata, R., & Yeh, C. (2016). The role of perception, interpretation, and decision making in the development of beginning teachers' competence. *ZDM – The International Journal on Mathematics Education*, 48(1-2), 153–165. https://doi.org/10.1007/s11858-015-0737-9

Santagata, R., Zannoni, C., & Stigler, J. W. (2007). The role of lesson analysis in pre-service teacher education: An empirical investigation of teacher learning from a virtual video-based field experience. *Journal of mathematics teacher education*, 10(2), 123–140.

Schochet, P. Z. (2008). *Technical Methods Report: Guidelines for Multiple Testing in Impact Evaluations* (NCEE 2008-4018). Washington DC: *National Center for Education Evaluation and Regional Assistance*.

Scherer, P., & Moser Opitz, E. (2012). *Fördern im Mathematikunterricht der Primarstufe*. Springer-Verlag.

Scheuer, N., Sinclair, A., Merlo de Rivas, S., & Tièche Christinat, C. (2000). Cuando ciento setenta y uno se escribe 10071: niños de 5 a 8 años produciendo numerales. *Infancia y aprendizaje*, 23(90), 31–50. DOI: https://doi.org/10.1174/021037000760087955

Schleppenbach, M., Flevares, L. M., Sims, L. M., & Perry, M. (2007). Teachers' responses to student mistakes in Chinese and US mathematics classrooms. *The elementary school journal, 108*(2), 131–147.

Schoenfeld, A. H. (2011). Toward professional development for teachers grounded in a theory of decision making. *ZDM, 43*(4), 457–469.

Schrader, F.-W. (2006). Diagnostische Kompetenz von Eltern und Lehrern. In D. H. Rost (Eds.), *Handwörterbuch Pädagogische Psychologie* (S. 68–71). Weinheim: Beltz.

Schrader, F. W. (2013). Diagnostische Kompetenz von Lehrpersonen. *Beiträge zur Lehrerinnen-und Lehrerbildung, 31*(2), 154–165.

Schrader, F. W., & Helmke, A. (1987). Diagnostische Kompetenz von Lehrern: Komponenten und Wirkungen. *Empirische Pädagogik, 1*(1), 27–52.

Schwarz, B. (2015). A study on professional competence of future teacher students as an example of a study using qualitative content analysis. In A. Bikner-Ahsbahs, C. Knipping & N. Presmeg (Eds.), *Approaches to qualitative research in mathematics education* (pp. 381–399). Dordrecht: Springer.

Schwarz, B., Wissmach, B. & Kaiser, G. (2008). "Last curves not quite correct": diagnostic competences of future teachers with regard to modelling and graphical representations. *ZDM – The International Journal on Mathematics Education, 40*(5), 777–790. doi: https://doi.org/10.1007/s11858-008-0158-0

Seifried, J., & Wuttke, E. (2010a). Student errors: How teachers diagnose them and how they respond to them. *Empirical research in vocational education and training, 2*(2), 147–162.

Seifried, J., & Wuttke, E. (2010b). Professionelle Fehlerkompetenz-Operationalisierung einer vernachlässigten Kompetenzfacette von (angehenden) Lehrkräften. *Wirtschaftspsychologie, 12*(4), 17–28.

Seifried, J., Wuttke, E., Türling, J. M., Krille, C., & Paul, O. (2015). Teachers' strategies for handling student errors–the contribution of teacher training programs. In M. Gartmeier, H. Gruber, T. Hascher & H. Heid (Eds.) *Fehler: Ihre Funktionen im Kontext individueller und gesellschaftlicher Entwicklung* (pp. 177–188). Münster: Waxmann.

Selter, C., Prediger, S., Nührenbörger, M., & Hußmann, S. (2012). Taking away and determining the difference—a longitudinal perspective on two models of subtraction and the inverse relation to addition. *Educational Studies in Mathematics, 79*(3), 389–408.

Selter, C. & Spiegel, H. (1997). *Wie Kinder rechnen.* Leipzig: Klett.

Sharpe, D. (2015). Chi-Square Test is Statistically Significant: Now What?. *Practical Assessment, Research, and Evaluation, 20*(8), 1–10.

Shavelson, R. J. (1983). Review of research on teachers' pedagogical judgments, plans, and decisions. *The Elementary school journal, 83*(4), 392–413.

Shavelson, R. J., Atwood, N., & Borko, H. (1977). Experiments on some factors contributing to teachers' pedagogical decisions. *Cambridge Journal of Education, 7*, 51–70.

Shavelson, R. J. & Stern, P. (1981). Research on Teachers' Pedagogical Thoughts, Judgements, Decisions and Behavior. *Review of Educational Research, 51*(4), 455–498.

Scherer, P., & Moser Opitz, E. (2012). *Fördern im Mathematikunterricht der Primarstufe.* Heidelberg: Springer-Spektrum.

Sherin, M. G., Jacobs, V. R., & Philipp, R. A. (2011). Situating the study of teacher noticing. In M. G. Sherin, V. R. Jacobs, & R. A. Philipp (Eds.), *Mathematics teacher noticing: Seeing through teachers' eyes* (pp. 3–13). New York: Routledge.

Sherin, M. G., Russ, R. S., & Colestock, A. A. (2011). Accessing mathematics teachers' in-the-moment noticing: Seeing through teachers' eyes. In M. G. Sherin, V. R. Jacobs, & R. A. Philipp (Eds.), *Mathematics teacher noticing: Seeing through teachers' eyes* (pp. 79–94). New York: Routledge.

Sherin, M., & van Es, E. (2005). Using video to support teachers' ability to notice classroom interactions. *Journal of technology and teacher education, 13*(3), 475–491.

Sheskin, D. J. (2000). *Handbook of parametric and nonparametric statistical procedures* (2nd ed). Boca Raton: Chapman & Hall/CRC.

Shulman, L. S. (1986). Those who understand: Knowledge growth in teaching. *Educational Researcher*, 15(2), 1–22.

Shulman, L. S. (1987). Knowledge and teaching: foundations of the new reform. *Harvard Educational Research, 57,* 1–22.

Smith, J. P., diSessa, A.A. & Roschelle, J. (1993). Misconceptions reconceived: A constructivist analysis of knowledge in transition. *The Journal of the Learning Sciences, 3*(2), 115–163.

Snow, R. (1972). A model teacher training system: an overview. *Research and Development Memorandum No. 92.* Stanford Center for Research and Development in Teaching.

Son, J. W. (2013). How preservice teachers interpret and respond to student errors: ratio and proportion in similar rectangles. *Educational studies in mathematics, 84*(1), 49–70.

Son, J. W., & Crespo, S. (2009). Prospective teachers' reasoning and response to a student's non-traditional strategy when dividing fractions. *Journal of Mathematics Teacher Education, 12*(4), 235–261.

Son, J. W., & Senk, S. L. (2010). How reform curricula in the USA and Korea present multiplication and division of fractions. *Educational Studies in Mathematics, 74*(2), 117–142.

Son, J. W., & Sinclair, N. (2010). How preservice teachers interpret and respond to student geometric errors. *School Science and Mathematics, 110*(1), 31–46.

Spinath, B. (2005). Akkuratheit der Einschätzung von Schülermerkmalen durch Lehrer und das Konstrukt der diagnostischen Kompetenz [Accuracy of teacher judgments on student characteristics and the construct of diagnostic competence]. *Zeitschrift für pädagogische Psychologie, 19*(1/2), 85–95.

Spychiger, M., Oser, F., Hascher, T. & Mahler, F. (1999). Entwicklung einer Fehlerkultur in der Schule. In W. Althof (Ed.) *Fehlerwelten: vom Fehlermachen und Lernen aus Fehlern* (pp. 43–70). Springer Fachmedien, Wiesbaden.

Stahnke, R., Schueler, S., & Roesken-Winter, B. (2016). Teachers' perception, interpretation, and decision-making: a systematic review of empirical mathematics education research. *ZDM – The International Journal on Mathematics Education, 48*(1-2), 1–27.

Star, J. R., & Strickland, S. K. (2008). Learning to observe: Using video to improve preservice mathematics teachers' ability to notice. *Journal of Mathematics teacher Education, 11*(2), 107–125.

Star, J. R., Lynch, K. H., & Perova, N. (2011). Using video to improve mathematics' teachers' abilities to attend to classroom features: A replication study. In M. G. Sherin, V. R. Jacobs, & R. A. Philipp (Eds.), *Mathematics teacher noticing: Seeing through teachers' eyes* (pp. 117–133). New York: Routledge.

Südkamp, A., Kaiser, J., & Möller, J. (2012). Accuracy of teachers' judgments of students' academic achievement: A meta-analysis. *Journal of educational psychology, 104*(3), 743.

Tatto, M. T., Schwille, J., Senk, S., Ingvarson, L., Peck, R., & Rowley, G. (2008). *Teacher Education and Development Study in Mathematics (TEDS-M): Policy, practice, and readiness to teach primary and secondary mathematics. Conceptual framework.* East Lansing, MI: Teacher Education and Development International Study Center, College of Education, Michigan State University.

Tekle, F. B., Gudicha, D. W., & Vermunt, J. K. (2016). Power analysis for the bootstrap likelihood ratio test for the number of classes in latent class models. *Advances in Data Analysis and Classification, 10*(2), 209–224.

Türling, J. M., Seifried, J., & Wuttke, E. (2012). Teachers' knowledge about domain specific student errors. *Learning from errors at school and at work, 1*, 95–110.

Van de Walle, J. A., Lovin, L. A. H., Karp, K. S., & Bay-Williams, J. M. (2014). *Teaching Student-centered Mathematics: Developmentally Appropriate Instruction for Grades 3–5* (Vol. 2). Pearson Higher Ed.

van Es, E. A. (2011). A framework for learning to notice student thinking. In M. G. Sherin, V. R. Jacobs, & R. A. Philipp (Eds.), *Mathematics teacher noticing: Seeing through teachers' eyes* (pp. 164–181). New York: Routledge.

van Es, E. A., Cashen, M., Barnhart, T., & Auger, A. (2017). Learning to notice mathematics instruction: Using video to develop preservice teachers' vision of ambitious pedagogy. *Cognition and Instruction, 35*(3), 165–187.

van Es, E. A., & Sherin, M. G. (2002). Learning to notice: Scaffolding new teachers' interpretations of classroom interactions. *Journal of technology and teacher education, 10*(4), 571–596.

van Es, E. A., & Sherin, M. G. (2006). How different video club designs support teachers in "learning to notice". *Journal of computing in teacher education, 22*(4), 125–135.

Vermunt, J.K. & Magidson, J. (2004). Latent class analysis. In: M.S. Lewis-Beck, A. Bryman, and T.F. Liao (Eds.), *The Sage Encyclopedia of Social Sciences Research Methods*, 549–553. Thousand Oaks, CA: Sage Publications.

Victor, A., Elsässer, A., Hommel, G., & Blettner, M. (2010). Judging a plethora of p-values: how to contend with the problem of multiple testing–part 10 of a series on evaluation of scientific publications. *Deutsches Arzteblatt international, 107*(4), 50–56. doi: https://doi.org/10.3238/arztebl.2010.0050

von Aufschnaiter, C. V., Cappell, J., Dübbelde, G., Ennemoser, M., Mayer, J., Stiensmeier-Pelster, J., Sträßer, R., & Wolgast, A. (2015). Diagnostische Kompetenz. Theoretische Überlegungen zu einem zentralen Konstrukt der Lehrerbildung. *Zeitschrift für Pädagogik, 61*(5), 738–758.

Watanabe, T. (2002). Representations in teaching and learning fractions. *Teaching Children Mathematics*, 8(8): 457–463.

Waterbury, G. T. (2019). Missing Data and the Rasch Model: The Effects of Missing Data Mechanisms on Item Parameter Estimation. *Journal of Applied Measurement, 20*(2), 154–166.

Weijters, B., Geuens, M., & Schillewaert, N. (2010). The stability of individual response styles. *Psychological methods, 15*(1), 96.

Weimer, H. (1925). *Psychologie der Fehler*. Leipzig: Julius Klinkhardt.

Weinert, F. E. (1999). *Konzepte der Kompetenz. Gutachten zum OECD-Projekt "Definition and Selection of Competencies: Theoretical and Conceptual Foundations (DeSeCo)."*. Neuchatel: Bundesamt für Statistik.

Weinert, F. E. (2001). Concept of competence: A conceptual clarification. In D. S. Rychen, & L. H. Saganik (Eds.), *Defining and selecting key competencies* (pp. 45–65). Seattle: Hogrefe & Huber.

Welch, B. (1951). On the comparison of several mean values: An alternative approach. *Biometrika, 38,* 330–336.

Wildgans-Lang, A., Scheuerer, S., Obersteiner, A., Fischer, F. & Reiss, K. (2020). Analyzing prospective mathematics teachers' diagnostic processes in a simulated environment. *ZDM – The International Journal on Mathematics Education, 52*(2), 241–254. https://doi.org/10.1007/s11858-020-01139-9

Wu, M., & Adams, R. (2007). *Applying the Rasch model to psycho-social measurement: A practical approach.* Melbourne: Educational Measurement Solutions.

Wu, M., Tam, H. P., & Jen, T. H. (2016). *Educational measurement for applied researchers. Theory into practice.* Singapore: Springer. https://doi.org/10.1007/978-981-10-3302-5_5.

Wuttke, E., & Seifried, J. (2013). Diagnostic competence of (prospective) teachers in vocational education: An analysis of error identification in accounting lessons. In K. Beck & O. Zlatkin-Troitschanskaia (Eds.) *From Diagnostics to Learning Success: Proceedings in Vocational Education and Training* (pp. 225–240). Rotterdam: Sense.

Yang, X., Kaiser, G., König, J., & Blömeke, S. (2020). Relationship between Chinese mathematics teachers' knowledge and their professional noticing. *International Journal of Science and Mathematics Education.* https://doi.org/10.1007/s10763-020-10089-3

Yang, F. M. & Kao, S. T. (2014). Item response theory for measurement validity. *Shanghai Archives of Psychiatry, 26*(3), 171–177.

Young, R. M., & O'Shea, T. (1981). Errors in children's subtraction. *Cognitive Science, 5*(2), 153–177.

Zahner, W., Velazquez, G., Moschkovich, J., Vahey, P., & Lara-Meloy, T. (2012). Mathematics teaching practices with technology that support conceptual understanding for Latino/a students. *The Journal of Mathematical Behavior, 31*(4), 431–446.

Printed in the United States
by Baker & Taylor Publisher Services